Physics for the Anaesthetic Viva

Physics for the Anaesthetic Viva

Aman Kalsi
ST7 Anaesthetist, University Hospital Southampton, Wessex School of Anaesthesia, Southampton, UK

Nikhail Balani
ST5 Anaesthetist, University Hospital Lewisham, South East School of Anaesthesia, London, UK

CAMBRIDGE
UNIVERSITY PRESS

University Printing House, Cambridge CB2 8BS, United Kingdom

Cambridge University Press is part of the University of Cambridge.

It furthers the University's mission by disseminating knowledge in the pursuit of education, learning and research at the highest international levels of excellence.

www.cambridge.org
Information on this title: www.cambridge.org/9781107498334

First published 2016

Printed in the United Kingdom by Clays, St Ives plc

A catalogue record for this publication is available from the British Library

Library of Congress Cataloguing in Publication data
Names: Kalsi, Aman, author. | Balani, Nikhail, author.
Title: Physics for the anaesthetic viva / Aman Kalsi, Nikhail Balani.
Description: Cambridge, United Kingdom ; New York : Cambridge University Press, 2016. | Includes bibliographical references and index.
Identifiers: LCCN 2015040113 | ISBN 9781107498334 (pbk.)
Subjects: | MESH: Anesthesiology. | Physics. | Anesthetics. | Equipment and Supplies. | Physical Phenomena. | Physicochemical Phenomena.
Classification: LCC RD81 | NLM WO 200 | DDC 617.9/6—dc23 LC record available at http://lccn.loc.gov/2015040113

ISBN 978-1-107-49833-4 Paperback

Contents

Section 7. Gas analysis 83

Section 8. Suction, CO_2 absorption and cleaning 99

Section 9. Radiology 107

Preface

The FRCA primary and final vivas are monumental tasks – yet extremely rewarding when achieved! They require both breadth and depth of knowledge in all areas of basic science and clinical practice.

The physics component of the viva forms a significant part, yet it is a topic that many trainees have not encountered since secondary school. For us, the months and weeks before the viva involved carrying around several kilograms of revision textbooks. This was not only impractical but exhausting – and costly in terms of physiotherapy!

With that in mind, we set out to create a physics revision aid that addresses commonly encountered topics in the viva. It is arranged to succinctly explain the basic science concepts, followed by a relevant clinical application of these. At the end of each chapter we have included some sample questions to help direct learning.

Our aim was to create a book that can be used on your commute to work or during one of your 'rarely' encountered anaesthetic coffee breaks. We hope that you find it complements your revision and wish you success in the examination.

We owe a world of thanks to our significant others, Jas Kalsi and Kim Saul, whose patience with us and assistance with proof reading have allowed this text to come to fruition.

Aman Kalsi and Nikhail Balani

Section 1 Mathematical concepts

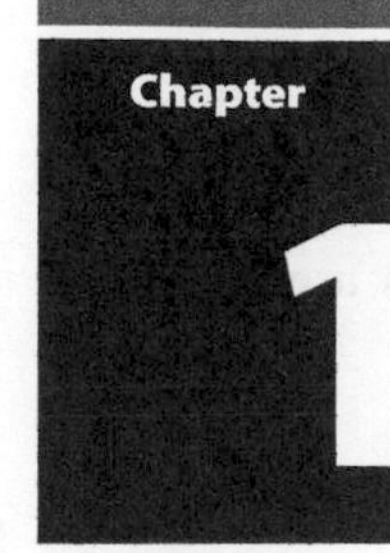

SI units

Basic science

Le Système International d'Unités (SI units) is an internationally agreed system of standardised units of measure. It comprises two classes of unit: base units and derived units.

Seven base units form the building blocks on which the system is based and they are assumed to be mutually independent.

Length has the unit **metre (m)**

- The length of the path travelled by light in vacuum during a time interval of 1/299792458 of a second.[1]

Mass has the unit **kilogram (kg)**

- The mass of the international prototype of the kilogram.[1]

Time has the unit **seconds (s)**

- The duration of 9192631770 periods of the radiation corresponding to the transition between the two hyperfine levels of the ground state of the cesium-133 atom.[1]

Electric current has the unit **ampere (A)**

- The constant current which, if maintained in two straight parallel conductors of infinite length, of negligible circular cross-section, and placed 1 metre apart in vacuum, would produce between these conductors a force equal to 2×10^{-7} newton per metre of length.[1]

Thermodynamic temperature has the unit **kelvin (K)**

- The fraction 1/273.16 of the thermodynamic temperature of the triple point of water.[1]

Luminous intensity has the unit **candela (cd)**

- The luminous intensity, in a given direction, of a source that emits monochromatic radiation of frequency 540×10^{12} hertz and that has a radiant intensity in that direction of (1/683) watt per steradian.[1]

Amount of substance has the unit **mole (mol)**

- The amount of substance of a system which contains as many elementary entities as there are atoms in 0.012 kilogram of carbon 12.[1]

There are a series of **SI unit prefixes** that describe decimal multiples and submultiples of the SI units. These range from yotta (10^{24}) through to yocto (10^{-24}). The prefixes milli (10^{-3}), micro (10^{-6}) and nano (10^{-9}) are commonly encountered in anaesthesia for drug dosing.

Applied science

What is a derived SI unit? Give some examples

SI derived units are products of mathematical manipulation of the base units, usually by division and multiplication.

Quantity	Name	Symbol
Area	Square metre	m^2
Volume	Cubic metre	m^3
Acceleration	Metre per second squared	$m \cdot s^{-2}$
Density	Kilogram per cubic metre	$kg \cdot m^{-3}$

There are derived SI units that have **special names and symbols.** For example, the derived SI unit of force is the newton (N), which is given by mass (kg) multiplied by acceleration ($m \cdot s^{-2}$). Therefore, the newton expressed in terms of base SI units is $kg \cdot m \cdot s^{-2}$. Some other examples encountered in anaesthesia are given below.

Quantity	Special name	Symbol	Derivation
Frequency	Hertz	Hz	s^{-1}
Pressure	Pascal	Pa	$m^{-1} \cdot kg \cdot s^{-2}$
Resistance	Ohm	Ω	$m^2 \cdot kg \cdot s^{-3} \cdot A^{-2}$

Questions

What is the SI system?
How many base SI units are there?
List and define each of the base SI units.
What prefixes are used with SI units? Why are they used?
Give some examples of derived SI units that have special names.

Reference

[1] Le Systéme International d'Unités (SI), The International System of Units (SI), 8th Edition. Bureau International des Poids et Mesures, Sèvres, France, 2006. Accessed at www.bipm.org.

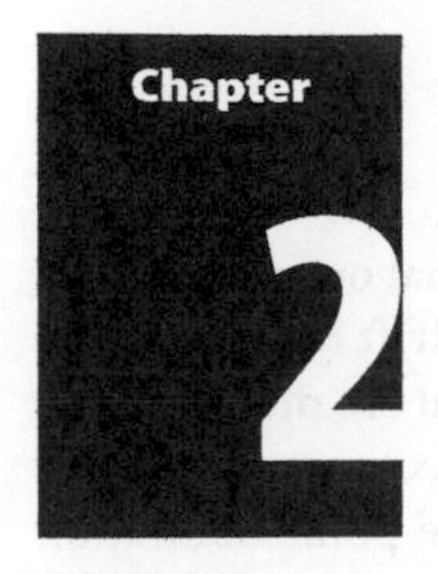

Measuring systems

Basic science

Calibration is important to ensure the readings of a measurement system are consistent with other measurements, accurate and reliable. Calibration establishes the correct relationship between input and output variables of a measuring system. A range of known inputs (accepted standards) is applied to a measuring system and the output at each variable is observed and used to construct a calibration curve. This process ensures the accuracy of measuring systems over the entire range of variables to be measured.

Accuracy, when applied to measuring devices, is the degree to which the measured value conforms to the correct value or standard. **Precision** describes the reproducibility of measurements, by assessing the degree to which repeated measurements of the same sample agree with their mean value. Measuring equipment should be accurate and precise to produce reliable and meaningful results.

The relationships between dependent and independent variables can be described as either linear or non-linear.

A **linear** relationship is graphically represented as a straight line. It is the extent to which any effect is exactly proportional to its cause. Resistors are an example of a linear device, where voltage is proportional to electric current if resistance is held constant.

In a **non-linear** relationship, the effect is not exactly proportional to the cause. The varying strength of the relationship between the variables gives rise to the curved line seen when graphically displaying non-linearity. Over some values, the relationship is stronger, producing a steep curve, while over others it is weaker, giving rise to a shallower curve. Relays and switches are examples of electronic devices that have a non-linear voltage–current relationship.

Hysteresis is a non-linear relationship that represents the **history dependence** of physical systems. The variable changes differently in response to either increasing or decreasing variations in the effect producing it. The compliance curves for the lung differ for inspiration and expiration. There is history dependence because the pressure–volume change varies depending on the degree of lung expansion. During inspiration, higher pressures are required to produce volume change due to tissue characteristics and recruitment of alveoli. In expiration, elastic recoil of tissues produces a different response.

Applied science

What is drift and how is it prevented?

Drift errors result from performance deviations of a measuring system that occur after calibration. The main types of drift error are offset and gain errors. **Offset drift** refers to those errors that are added to the measured value and **gain errors** are those that are multiplied by the measured value. Drift errors most commonly occur as a result of the effects of temperature on measuring devices and can be reduced by frequent calibration. Single-point calibration will correct offset errors, but a minimum of two-point calibration is required to correct gain errors.

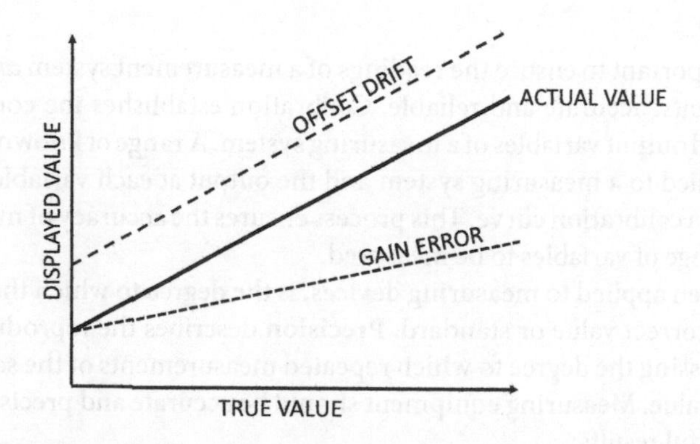

Static calibration applies a known value to a measuring system and observes the output. The values of the variables involved do not change during the measurement process (i.e. they are static). The magnitudes of the input and measured output are important.

Dynamic calibration is used in addition to static calibration when time-dependent variables, which change in magnitude and frequency with time, are to be measured. The relationship between the dynamic input and output of a measuring system depends on the time-dependent content of the input signal. For dynamic calibration, an input of known dynamic behaviour, such as a sinusoidal signal or a step change, is used to assess the system output.

Questions

What are linearity and non-linearity?
What is hysteresis?
What is the difference between static and dynamic and calibration?
What are drift errors?

Chapter 3

Simple mechanics: mass, force and pressure

Basic science

Mass is the amount of matter contained in an object, it is measured in kilograms (kg). **Force** is an interaction that changes an object's state of rest or motion, measured in newtons (N).

$$\textit{Force}\ (\mathrm{N}) = \textit{Mass}\ (\mathrm{kg}) \times \textit{Acceleration}\ (\mathrm{m{\cdot}s^{-1}})$$

Weight refers to the force exerted by a mass under the acceleration of gravity, also measured in newtons (N).

Pressure is force per unit area and has the SI unit pascal (Pa). The same force will generate a higher pressure over a smaller surface area than a larger one. For example, the same force applied to a 2-ml and a 50-ml syringe will generate a higher pressure in the 2-ml syringe.

$$\textit{Force}\ (\mathrm{N}) = \textit{Pressure}\ (\mathrm{Pa}) \times \textit{Area}\ (\mathrm{m^2})$$

In clinical practice, other units are also commonly used when referring to pressure, such as mmHg and atmosphere.

$$1\ \text{atmosphere} = 101.3\ \text{kPa} = 1.01\ \text{barr} = 760\ \text{mmHg} = 760\ \text{torr}$$

Gases exert a pressure resulting from the force generated by random collisions between gas molecules and surfaces. The greater the density of gas molecules, the greater the number of collisions; therefore, the higher the pressure exerted. In the same manner, the air molecules around us also exert a pressure known as **atmospheric pressure**. At sea level, atmospheric pressure is 101.3 kPa, which is the pressure exerted on surfaces by the weight of air molecules. Air density decreases with altitude due to the reduced effect of gravity.

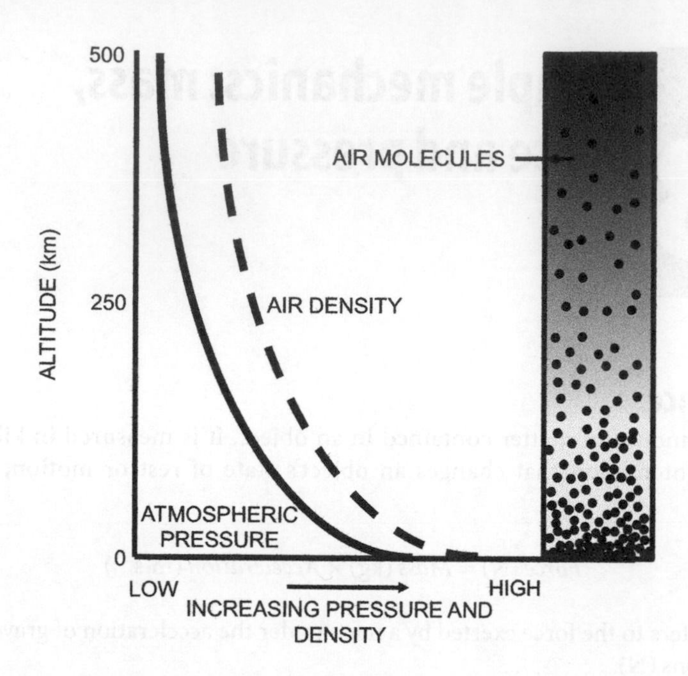

Gauge pressure is measured pressure relative to atmospheric pressure. **Absolute pressure** is pressure relative to a vacuum.

Absolute Pressure = Gauge Pressure + Atmospheric Pressure

Manometers measure gauge pressure. They consist of a liquid reservoir connected to a vertical glass column of liquid that is open to the atmosphere. An unknown pressure (to be measured) exerts a force on the reservoir, which forces the liquid to rise up the column. As the liquid column rises, it exerts a greater downwards force, until its weight equals the force of the pressure being measured. At this point the height of the column can be measured in mmHg or cmH_2O and represents the gauge pressure of the unknown pressure.

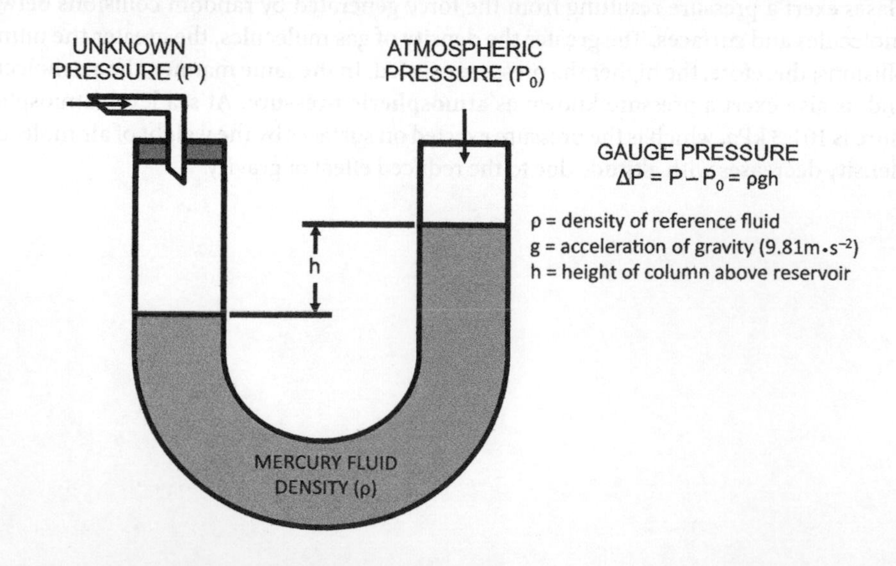

A **barometer** is used to measure atmospheric pressure. It has a similar design to the manometer, but has a sealed top creating a **toricellian vacuum** above the column of liquid. This is not a true vacuum, as it contains the saturated vapour pressure of the liquid in the column, e.g. mercury. If the toricellian vacuum were to become contaminated by another fluid, this would cause the level of the fluid column to fall by the vapour pressure exerted by the contaminant.

Applied science

How do we measure gas pressures in anaesthesia?

Gas pressures are most commonly measured using **anaeroid gauges** (anaeroid meaning 'without fluid') in the clinical setting. However, it is possible to use manometers or electronic strain gauges.

Bourdon gauges comprise a coiled, thin-walled, flattened tube with a closed end. An unknown pressure applied to the open end of the tube raises the pressure within, causing it to expand into a cylindrical shape and uncoil. This moves a pointer over a pressure scale, which has been calibrated against known pressures. In this way, the unknown pressure can be measured.

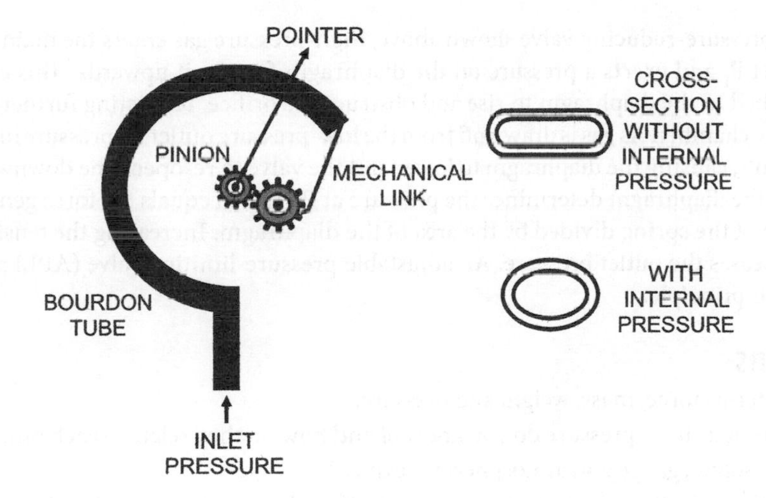

Diaphragm gauges consist of two chambers separated by a flexible ceramic membrane. A gas of unknown pressure introduced into a chamber causes deflection of the membrane towards the second chamber, which is usually open to atmosphere. The degree of deformation of the membrane is proportional to the pressure difference between the gases, and can be measured using optical techniques.

Bellows gauges are similar to bourdon gauges, in that an unknown pressure introduced into an elastic chamber causes expansion. The amount of expansion is measured and related to a pressure change on a scale.

How do single-stage pressure-reducing valves work?

Pressure-reducing valves can be found between gas cylinders and anaesthetic machines, allowing pressures to be reduced to a safe level for the machine to cope with. Some anaesthetic machines contain pressure-reducing valves between the pipeline supply and the machine to protect against fluctuations in pipeline pressures.

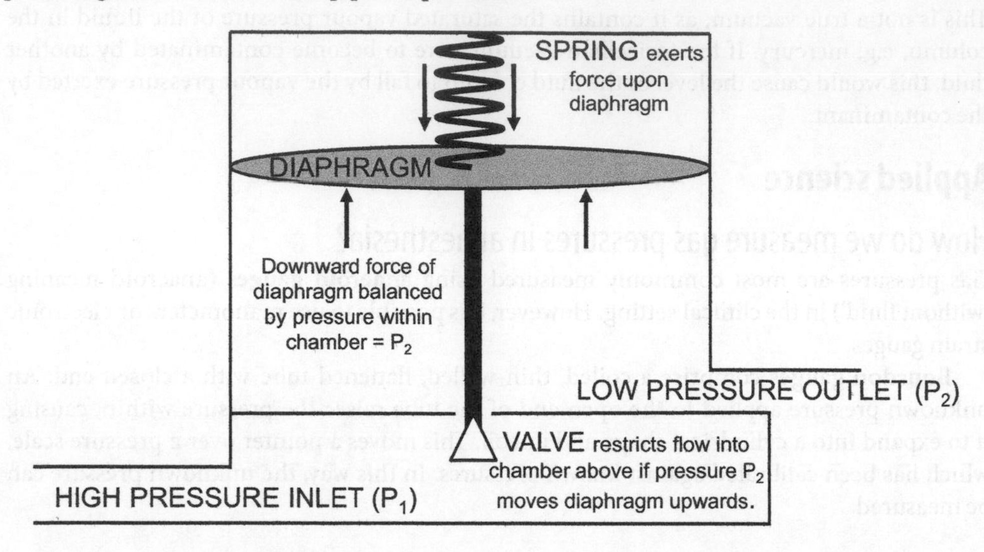

In the pressure-reducing valve shown above, high-pressure gas enters the main chamber via the inlet P_1 and exerts a pressure on the diaphragm forcing it upwards. This causes the valve attached to the diaphragm to rise and obstruct the orifice, restricting further gas from entering the chamber. As gas is drawn off from the low-pressure outlet P_2, pressure in the main chamber falls, causing the diaphragm to lower and the valve to re-open. The downward force exerted by the diaphragm determines the pressure at P_2, which equals the force generated by the tension of the spring divided by the area of the diaphragm. Increasing the tension of the spring increases the outlet pressure. An adjustable pressure-limiting valve (APL) also functions on this principle.

Questions

Define the terms force, mass, weight and pressure.
What different units of pressure do you know of and how do they relate to each other?
What are absolute, gauge and atmospheric pressure?
What is in the toricellian vacuum? What happens if isoflurane were placed in the vacuum?
Give a clinical example of the pressure, force and area relationship.

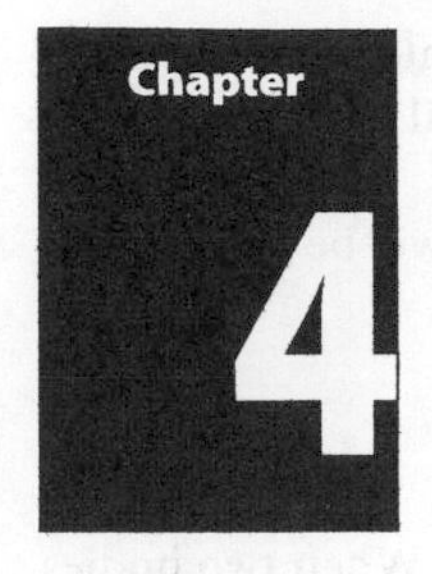

Simple mechanics: work, energy and power

Basic science

Work occurs when a force applied to an object causes displacement in the same direction as the force. It is measured in joules.

$$Work\ (\mathrm{J}) = Force\ (\mathrm{N}) \times Distance\ (\mathrm{m})$$

Energy is the capacity of a physical system to perform work. The **'conservation of energy principle'** states that energy cannot be created or destroyed, only transformed.

When considering mechanical work, the main classes of energy are potential and kinetic, although there are several subtypes of energy including heat, electric, nuclear, chemical and sound, all of which have the capacity to perform work.

Kinetic energy is the energy possessed by an object in motion. Kinetic energy is measured in joules.

$$Kinetic\ Energy\ (\mathrm{J}) = \frac{1}{2} \times Mass\ (\mathrm{kg}) \times Velocity^2\ (\mathrm{m \cdot s^{-1}})$$

Potential energy is the energy stored in an object due to its spatial position, and can be either gravitational or elastic.

Gravitational potential energy is the energy contained in an object that is at a specific location within a gravitational field. For example, if an object were released from a height, the force of gravity would pull the object towards Earth (performing work), and its potential energy would be transformed to kinetic energy.

Elastic potential energy occurs when certain materials such as a springs or elastic tissues (e.g. in our lungs) are stretched by a force. When the force is released, the recoil of the tissues to their original shape causes the elastic energy to be converted to kinetic energy.

Power is the rate of work done, and is measured in watts (W).

$$Power\ (\mathrm{W}) = \frac{Work\ (\mathrm{J})}{Time\ (\mathrm{s})}$$

Applied science

Describe Newton's laws of motion

Newton's laws describe the relationship between forces acting on a body and the subsequent motion produced.

First law: An object in a state of uniform motion remains in that state unless acted upon by an unbalanced force. Likewise, a motionless object will remain at rest until an external force acts upon it.

Second law: When a force acts upon a mass, the resulting acceleration will be in the direction of the force.

$$Force\ (\mathrm{N}) = Mass\ (\mathrm{kg}) \times Acceleration\ (\mathrm{m{\cdot}s^{-1}})$$

Larger objects require greater force to produce the same acceleration.

Third law: For every action, there is an equal and opposite reaction. When two bodies interact by exerting force on each other, these forces are equal in magnitude, but opposite in direction. For example, when a person walks, their foot exerts a force on the floor, and simultaneously the floor exerts an equal opposite force on their foot.

Questions

Define work, energy and power.
What is energy? What different types of energy are there?
Give examples of energy transfer from one type to another.
What is work?
What is kinetic energy?
What is the relationship between work and power?

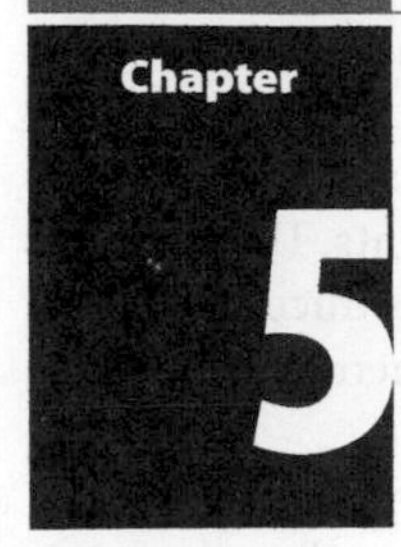

Heat and temperature

Basic science

Heat is a form of energy that is transferred between objects of differing temperatures. It is measured in joules.

Temperature is a measure of the average kinetic energy of the random movement of molecules within a substance. It quantifies how hot or cold a substance is, and gives an indication of how likely a substance is to transfer heat energy. Hot objects contain molecules with a higher average kinetic energy and will transfer this energy to cooler objects. The rate of energy transfer is dependent on the temperature gradient and thermal conductivity of the objects involved.

The SI unit for temperature is the **kelvin** (K), which is defined as 1/273.16 of the thermodynamic temperature of the triple point of water. The **triple point of water** is where ice, water and water vapour exist in equilibrium at a pressure of 611.2 pascals. Other commonly used scales are the Celsius and Fahrenheit.

There are four laws governing thermodynamics that describe how heat energy and other forms of energy are related.

The **zeroth law of thermodynamics** states that if two thermodynamic systems are each in thermal equilibrium with a third system, they are all in thermal equilibrium with each other.

The **first law of thermodynamics** states that in a closed system the total quantity of energy remains constant, as energy cannot be created or destroyed. Shivering generates heat from kinetic energy, which is generated from chemical energy within muscles. This is an example of the **mechanical equivalent of heat concept**.

The **second law of thermodynamics** states that the entropy of the universe (as a closed system) will increase over time. **Entropy** is a measure of disorder within a closed system. This is why heat is transferred from hotter to colder objects and not the other way round.

In the **third law of thermodynamics**, the entropy of a perfectly crystalline compound is zero at absolute zero. At **absolute zero** (0 K or –273.15°C), no thermal energy exists.

Applied science

Describe how heat may be lost from the patient intraoperatively, and how these losses can be minimised

Radiation is the transfer of heat via the emission of infrared waves, which accounts for roughly 40% of heat loss during anaesthesia. Drug-induced vasodilatation during anaesthesia increases radiative losses. Reflective blankets will return some of this radiant heat to the patient, and radiant heaters can be used to actively warm the patient.

Convection is the mass movement of liquid or gas from a hotter region to a cooler one. Hotter fluids rise because they are less dense, which creates convection currents. This accounts for approximately 30% of intraoperative heat loss. Clothing and blankets reduce convective losses by trapping warmed pockets of air around the patient, preventing them from moving away. Forced air warmers and heated fluids actively warm by convection.

Evaporation requires the latent heat of evaporation, which is provided by the patient, thus cooling them. Evaporative losses account for around 20% of total heat loss and can be minimised by covering open body cavities and wet surfaces, and increasing ambient humidity.

Evaporation from the respiratory tract and warming of fresh gas accounts for approximately 10% of heat loss. This can be minimised through the use of heat and moisture exchange filters, circle breathing systems, and by warming and humidifying inspired gases.

Conductive losses involve the flow of heat through solid matter via direct contact. They are only minimally responsible for heat loss during anaesthesia.

Questions

Define heat and temperature.
What is absolute zero?
Define the triple point of water.
Define the laws of thermodynamics.
What scales of temperature do you commonly use?
Explain the mechanisms of heat transfer.

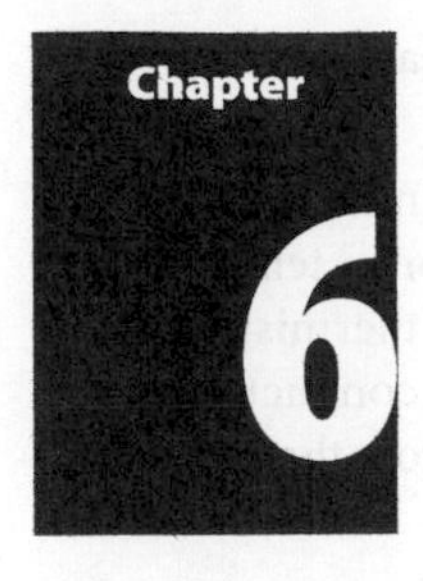

Temperature measurement

Basic science

Thermometric properties are predictable changes in physical properties of a substance that occur with alterations in temperature. They can be used in temperature measuring devices. They comprise the following.

- Change in volume of a liquid or solid.
- Change in pressure of a fixed mass of gas.
- Change in wavelength of light energy emitted.
- Change in electrical resistance and electromotive force.

Non-electrical devices

Liquid thermometers consist of a glass bulb of liquid mercury or alcohol connected to a vacuum tube. Heating causes liquid expansion along the tube, allowing temperature to be read from a scale.

Bourdon gauge thermometers consist of a fixed-volume reservoir of gas connected to a pressure gauge. Heating the gas causes pressure within the container to rise proportionally, which can be read off the bourdon gauge scale.

Bimetallic strip thermometers consist of strips of two dissimilar metals, such as steel and copper, joined together in a coil. Copper has a greater thermal expansion coefficient than steel. As temperature rises, copper expands at a greater rate, causing the strip to bend, which moves an attached pointer over a temperature scale.

Liquid crystal thermometers consist of layers of phospholipid crystals sandwiched between sheets of plastic. Each layer is orientated to lie at a different angle to the next, forming a twisted 'nematic' phase, which reflects light of a specific wavelength. Temperature changes alter the spacing of crystals, leading to the reflection of different wavelengths of light. This results in a colour change of the device.

Infrared thermometers detect wavelengths of thermal radiation emitted by objects and convert this into a temperature reading. They consist of a lens that focuses infrared radiation onto a pyroelectric crystal detector. This creates an electrical charge proportional to the wavelength of emitted radiation, which can be used to determine temperature.

Electrical devices

Resistance wires made of platinum are used with a Wheatstone bridge circuit. Heating causes agitation of molecules within the wire, which impedes the flow of electrons.

Therefore, as temperature rises, the electrical resistance of the wire increases. The Wheatstone bridge circuit is used to detect the change in resistance.

Thermistors are semi-conductors composed of a bead of metal oxide of nickel, iron, copper, manganese or cobalt. Their resistance falls in a non-linear fashion as temperature increases. This is because the electrons in the outermost shell of the thermistor gain energy, allowing them to break free from their valance band to enter the conductance band of the molecule they are associated with. Here they move freely across the structure to conduct electrical charge.

Thermocouples consist of a junction between two dissimilar metals such as copper and constantan (a copper and nickel alloy), which have differing thermal coefficients of expansion. When part of a metal conductor is heated, it expands and its density of electrons falls. At a junction between dissimilar metals a concentration gradient for electrons across the junction occurs. The potential difference created is known as the **Seebeck effect**. A reference junction at a fixed temperature connected to a voltmeter measures the potential difference created across the measuring junction, which is proportional to the temperature difference between the two junctions.

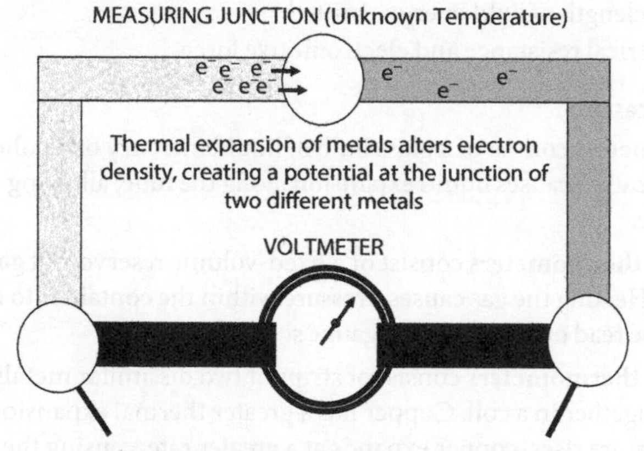

Applied science

Describe the applications of non-electrical thermometers

Liquid thermometers containing mercury are effective from −39°C to 250°C, while alcohol thermometers are effective from −117°C to 78°C. They are commonly used to measure temperature sublingually, rectally or within limb skin folds (e.g. axilla and groin). These devices are easily portable, cheap to produce and reusable. However, they are fragile and easily broken. In the case of mercury devices there is the added risk of mercury poisoning following damage. They have a slow response time of 2–3 minutes, and there is a risk of cross infection if not disinfected following use.

Liquid crystal thermometers are effective from 0°C to 120°C. Clinically they are used to measure temperature on the skin of the forehead. They are inexpensive, disposable, accurate to within 0.5°C, and have a relatively rapid response time of 1 minute. However, as they are placed on skin they may fail to reflect core temperature accurately in situations of peripheral vasoconstriction. To remain compact for clinical use, individual thermometers tend to only cover a small range of temperatures from 32°C to 40°C and are therefore unsuitable at extremes of physiological temperature.

Bourdon gauge thermometers are effective from –270°C to 1500°C. They are effective over a wide range of temperatures, cheap and robust. However, they are not very accurate and are prone to calibration errors.

Bimetallic strip thermometers are effective from –40°C to 500°C. They are simple, accurate and robust. However, they are relatively insensitive to small temperature changes and are not used to measure temperature in clinical practice. They do serve a role in temperature compensation in devices such as vaporisers.

Describe the applications of electrical thermometers

Infrared thermometers are effective from –50°C to 1050°C. They are commonly used to measure temperature from the tympanic membrane or skin of the forehead. They have a rapid response time of 1 second and are non-invasive. Their accuracy varies from 0.5°C to 1°C, and they can be affected by ambient radiation, e.g. from the user of the device, or wax in the external auditory canal. Improper use can lead to trauma of the ear canal and tympanic membrane.

Resistance wires are effective over the range –180°C to 1150°C. They are very accurate, precise and not prone to drift. Resistance wires are sensitive enough to measure temperature changes as small as 0.0001°C. They are large, fragile devices with slow response times of up to 50 s, and as a result are not often used in clinical practice.

Thermistors are effective from –90°C to 130°C. They are compact devices that have a rapid response time; for this reason, they are used in pulmonary artery floatation catheters and other electronic temperature probes. Their resistance changes over time, leading to drift error. They tend to be single-use devices because exposure to high temperature affects their calibration and may damage the device.

Thermocouples are effective over the range –250°C to 1150°C. They can be made compact and placed in catheters, probes and hypodermic needles. Thermocouples are relatively cheap and accurate, with rapid response times of 0.1 ms. However, they are prone to electrical noise and they require a second thermometer to measure the temperature at the reference junction.

Questions

What are non-electrical methods of temperature measurement?
What are electrical methods of temperature measurement?
What are the advantages and disadvantages of each method?
What sites can be used clinically and what are the advantages and disadvantages of each?
What is the Seebeck effect?

Chapter 7

Latent heat

Basic science

Heat capacity is the amount of heat energy required to raise a given object by one kelvin, and has the unit $J{\cdot}K^{-1}$.

Specific heat capacity is the amount of heat energy required to raise one kilogram of a substance by one kelvin, without causing a change in state. It is measured in $J{\cdot}kg^{-1}{\cdot}K^{-1}$.

Solids and liquids tend to have higher specific heat capacities than gases. For example, the specific heat capacity for water is approximately four times higher than for air (4160 versus 998 $J{\cdot}kg^{-1}{\cdot}K^{-1}$).

All molecular and atomic particles exert attractive and repulsive forces on each other known as intermolecular forces of attraction. In a **solid** these particles oscillate about a fixed position, which maintains a given shape and volume. The addition of heat increases the kinetic energy of the molecules, until eventually the particles gain sufficient energy to overcome the intermolecular forces of attraction. When this happens the solid loses its fixed structure to become a liquid. This is known as the **melting point**.

A **liquid** is a nearly incompressible fluid that conforms to the shape of its container and has a constant volume. At its **boiling point** the molecules have sufficient energy to break free of the intermolecular forces of attraction and enter the gaseous phase.

When heat is applied to matter, temperature increases until the melting or boiling point is reached. At these points the addition of further heat energy is used to change the state of matter from solid to liquid and from liquid to gas. This does not cause a change in temperature. The energy required at these points is referred to as **latent heat of fusion** and **latent heat of vaporisation**, respectively.

Specific latent heat is the heat required to convert one kilogram of a substance from one phase to another at a given temperature. As temperature increases, the amount of additional energy required to overcome the intermolecular forces of attraction falls until the **critical temperature** of a substance is reached. At this point the specific latent heat is zero, as no further energy is required to complete the change in state of the substance.

Applied science

Where are the principles of heat capacity and latent heat applicable to anaesthesia?

Variable bypass vaporisers function by passing a small amount of fresh gas through the vaporising chamber, which is fully saturated with anaesthetic vapour. This removes vapour from the chamber. Further vaporisation from the anaesthetic liquid must occur to replace the vapour removed, which requires energy from the latent heat of vaporisation. This cools the remaining liquid, reducing the saturated vapour pressure and thus the concentration of anaesthetic vapour delivered, resulting in an unreliable device. Temperature compensation features help to overcome this problem; a copper heat sink placed around the vaporising chamber is one such example. Copper has a high heat capacity and donates energy required for latent heat of vaporisation, maintaining a stable temperature and reliable delivery of anaesthetic agent.

Evaporation of sweat is another example. It requires the latent heat of vaporisation, which is provided by the skin's surface, exerting a cooling effect upon the body. Evaporation from open body cavities can be a cause of significant heat loss from patients while under anaesthesia.

These principles are also applicable to blood transfusion. Blood is stored at 5°C and has a specific heat capacity of 3.5 $kJ \cdot kg^{-1} \cdot K^{-1}$. If cold blood were transfused into a patient without pre-warming, the heat energy required to warm the blood to body temperature would need to be supplied by the patient, which would have a significant cooling effect.

Questions

Define solids, liquids and gases.
Define boiling and freezing points.
Define latent heat.
What is specific latent heat?
What is heat capacity? How does it differ from specific heat capacity?

Chapter 8

Humidity

Basic science

Absolute humidity is the mass of water vapour in a given volume of gas at a specified temperature and pressure. It is measured as mass per unit volume, for example $g{\cdot}m^{-3}$ or $mg{\cdot}l^{-1}$.

Relative humidity is the ratio of the mass of water vapour present in a volume of gas to the mass required to fully saturate that same volume. It is expressed as a percentage.

Cooling or compressing a gas has no effect on absolute humidity, but causes a rise in relative humidity because cooler air holds less water vapour. Where the relative humidity reaches 100%, condensation occurs. This is known as the **dew point**.

Hygrometers are used to measure humidity. Depending on the device, it will either measure absolute or relative humidity.

The **hair hygrometer** consists of human or animal hair attached to a spring. As humidity rises, the length of the hair increases, which subsequently moves a pointer over a scale.

The **psychrometer (wet and dry bulb hygrometer)** consists of a wet wick enclosing a thermometer. Evaporation of water from the wick cools the thermometer. The rate of evaporation is directly related to the relative humidity. Relative humidity is determined by comparing the temperature of the wet thermometer to ambient temperature.

Regnault's hygrometer bubbles air through a silver tube containing ether, which evaporates and cools the tube. Condensation on the tube indicates the dew point. Measuring the temperature at this point allows determination of the absolute humidity from reference tables.

The **gravimetric hygrometer** measures the mass of an air sample and compares this to the mass of an equal volume of dry air. It is very accurate and measures absolute humidity.

Humidity transducers measure the change in resistance or capacitance of a material such as lithium chloride when it absorbs moisture, and translate this into humidity.

Other methods include mass spectrometry, Raman spectroscopy and absorption spectrophotometry.

Applied science

How are inspired gases humidified?

Passive systems include the **heat–moisture exchanger (HME)**, which consists of hygroscopic foam or a paper membrane containing calcium or lithium chloride. Expired gases cool as they pass through the HME, and water vapour condenses onto it. During the subsequent inspiration, stored heat and water are transferred to the inspired gas, providing humidification.

HMEs are light, inexpensive and can achieve a relative humidity of up to 60%–70%. They commonly come with an incorporated bacterial filter.

Bubbling gas through a **cold water bath** produces humidification of up to 30% relative humidity. During continuous use, evaporation cools the water, which reduces its humidifying capability. **Soda lime's** exothermic reaction produces heat and water, which can achieve a relative humidity of 60%–70% when used in a circle system.

Active systems include **hot water baths,** where gas is passed through water heated to 60°C. The gas cools as it passes along the breathing circuit, providing fully saturated gas at body temperature. Monitoring temperature at the patient end with a thermistor to regulate the heater reduces the risk of scalding and hyperthermia.

Gas-driven nebulisers force gas at high pressure across a water-filled tube. The Venturi effect entrains water droplets into the gas stream, which collide with an anvil, producing fine droplets of 2–5 μm. An **ultrasonic nebuliser** drops water onto a vibrating surface, producing droplets of 0.5–2 μm. Ideal droplet size is around 1 μm, as larger droplets are deposited in the trachea, while smaller ones deposit in alveoli, obstructing them. Nebulisers produce a relative humidity up to and in excess of 100%. A relative humidity over 100% is known as **supersaturation**, and is where microdroplets of water become suspended in the gas.

Questions

What is absolute humidity? How does relative humidity differ?
What is the dew point?
How does a Regnault's hygrometer function?
Describe the function of an HME.

Section 3 Gases, vapours and liquids

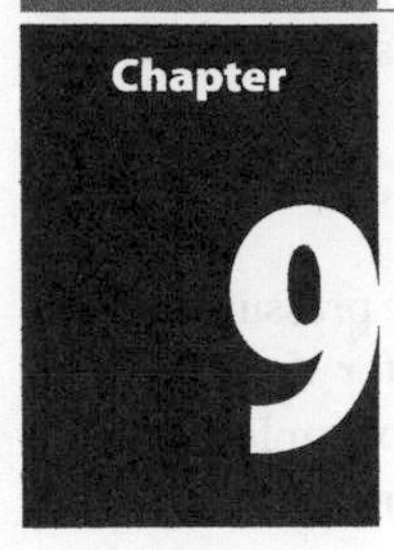

Physics of gases

Basic science

A **gas** is one of three fundamental states of matter, which has neither a definite shape nor size, and whose atoms possess perfect molecular mobility.

The **kinetic theory of gases** postulates that gases are composed of vast numbers of molecules separated by large distances. The molecules move and collide randomly with each other and the walls of the container, but otherwise do not interact. These collisions are perfectly elastic, meaning that no energy is transferred. The kinetic energy of gas particles is proportional to their temperature.

The **ideal gas equation** relates the pressure, temperature and volume of an ideal gas. It encompasses the three perfect gas laws described below.

$$PV = nRT$$

where P is absolute pressure (Pa), V is volume (m^3), n is amount of substance (mol), R is the universal gas constant ($8.314\ J \cdot K^{-1} \cdot mol^{-1}$), and T is temperature (K).

Boyle's law: at a constant temperature, the volume of an ideal gas is indirectly proportional to pressure.

Charles's law: at a constant pressure, the volume of an ideal gas is directly proportional to absolute temperature.

Gay-Lussac's law: at a constant volume, the pressure of an ideal gas is directly proportional to absolute temperature.

Other gas laws of relevance are as follows.

Avogadro's hypothesis: at a constant temperature and pressure, all gases of the same volume contain an equal number of molecules.

Dalton's law: the pressure exerted by a mixture of gases is the sum of the partial pressures of its constituents.

Henry's law: at a constant temperature, the amount of gas dissolved in a given volume of liquid is directly proportional to the partial pressure of that gas in equilibrium with the liquid.

Applied science

Discuss the anaesthetic relevance of Boyle's and Henry's laws

According to Boyle's law, at a constant temperature the product of pressure and volume is a constant. This means that as one decreases, the other must increase. As gas is released from a cylinder, its pressure falls to atmospheric pressure and its volume increases. The following equation can be used to describe this situation.

$$P_1V_1 = P_2V_2$$

P_1V_1 relates to the pressure and volume in the cylinder and P_2V_2 relates to the pressure and volume at atmospheric pressure. Rearranging and solving the above equation for V_2 will give the volume of gas that the cylinder will provide at atmospheric pressure. For example, oxygen is stored at 13 800 kPa (absolute pressure) in gas cylinders. If the internal volume of the cylinder is 10 litres, the volume this cylinder will provide at atmospheric pressure:

$$13\,800 \times 10 = 100 \times V_2$$

$V_2 = 1380$ litres. However, 10 litres will remain within the cylinder, so 1370 litres will be usable at atmospheric pressure.

Henry's law can be used to show that the amount of oxygen dissolved in blood is proportional to the partial pressure of oxygen in the alveolus. The amount of dissolved oxygen carried in blood is 0.023 $ml \cdot dl^{-1} \cdot kPa^{-1}$. At atmospheric pressure, this accounts for a very small and insignificant fraction of oxygen delivery. However, under hyperbaric conditions, the dissolved fraction increases and becomes a more significant source of oxygen delivery to tissues.

Questions

What is a gas?
Define the perfect gas laws and describe their relation to the ideal gas equation.
What is Avogadro's hypothesis?
What other gas laws are you aware of?
How can Boyle's law be used to calculate the amount of oxygen in a size E cylinder?

Chapter 10 Vapours

Basic science

A **vapour** is a substance in its gaseous phase, below its critical temperature. It can be liquefied by the addition of pressure alone. A **gas** is a substance above its critical temperature.

Critical temperature is the temperature above which a substance cannot be liquefied by pressure alone. The **critical pressure** is the pressure required to liquefy a vapour at its critical temperature.

At the surface of liquids, a vapour phase exists. This arises when molecules acquire sufficient energy to overcome the intermolecular forces of attraction and leave the liquid state. In a closed container, vapour will eventually reach dynamic equilibrium with the liquid, as molecules are continuously leaving and returning to the liquid at the same rate. At this point, the vapour is saturated and will exert a pressure known as the **saturated vapour pressure** (SVP). The pressure exerted is the result of random collisions of vapour molecules with each other and the walls of the container.

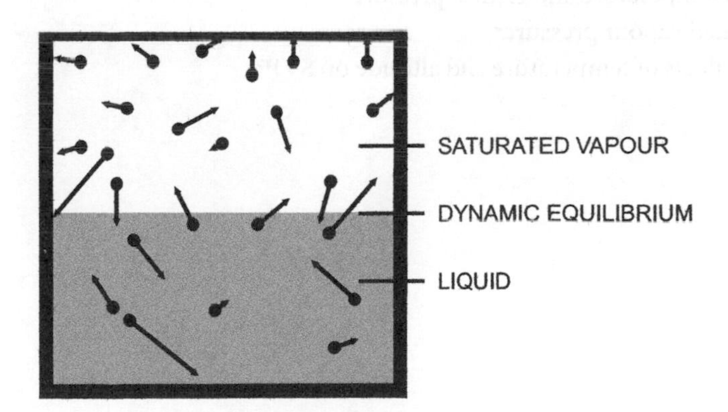

As a liquid's temperature rises, more molecules gain sufficient energy to leave the liquid and enter the vapour phase. This results in an increase in SVP. SVP continues to rise with temperature, until the liquid boils. At the boiling point, SVP equals atmospheric pressure.

The **latent heat of vaporisation** is the heat energy required to change the state of a substance from liquid to vapour. If vapour is removed from the liquid surface there will be a net loss of energy from the liquid, causing it to cool.

Applied science

How is the function of vaporisers affected by altitude?

Vaporisers are designed to deliver an accurate and precise concentration of a volatile agent. In plenum vaporisers, gas leaving the vaporising chamber is fully saturated with anaesthetic vapour. A concentration dial controls a valve within the bypass chamber, which enables the user to change the concentration of volatile delivered to the patient by altering the splitting ratio.

With increasing altitude, atmospheric pressure falls in a non-linear manner. However, vapour pressure does not change with altitude. Therefore, a vaporiser used at altitude will deliver a higher percentage concentration of volatile than is selected, but the partial pressure of the volatile agent remains the same. It is the partial pressure of the volatile anaesthetic agent that is important in determining its anaesthetic effect and not the concentration. Because the partial pressure remains unchanged at altitude, the desired isoflurane concentration can be set to the same value as it would at sea level.

Atmospheric pressure is 760 mmHg at sea level, which falls to approximately half at an altitude of 5500 m. If a plenum isoflurane vaporiser were set to deliver 3% isoflurane concentration, at 5500 m above sea level it would actually deliver a 6% concentration. In both instances, the partial pressure would be unchanged at 22.8 mmHg (3% of 760 mmHg = 6% of 380 mmHg = 22.8 mmHg).

Questions

What is the difference between a vapour and a gas?
Define critical temperature and critical pressure.
What is saturated vapour pressure?
What are the effects of temperature and altitude on SVP?

Chapter 11

Vaporisers

Basic science

Anaesthetic vaporisers are devices that are designed to deliver a safe, reliable and precise concentration of anaesthetic agents.

Vaporisers can be **variable bypass** (draw-over and plenum) or **measured flow** (desflurane and direct injection).

Variable bypass vaporisers consist of a vapour inlet port, concentration control dial, bypass chamber, a vaporising chamber and an outlet port. The fresh gas flow is split into two streams, the **bypass flow** and the **vaporiser flow.** Gas entering the vaporising chamber picks up anaesthetic vapour before it rejoins the bypass flow. The concentration dial adjusts the final concentration of delivered anaesthetic agent by varying the **splitting ratio**, which is the ratio of vaporiser flow to bypass flow.

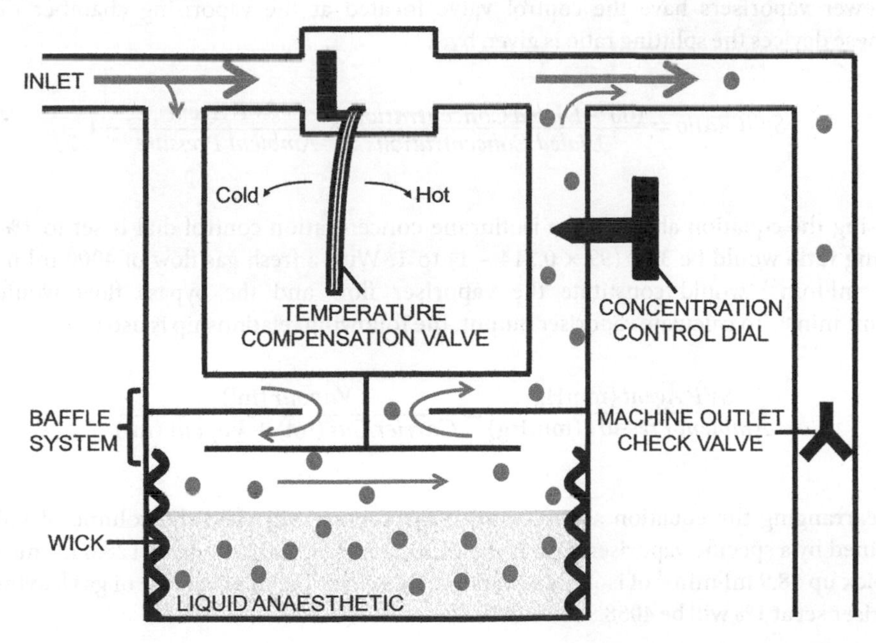

At room temperature and one atmosphere pressure, most anaesthetic agents exist in their liquid form. The vaporising chamber is fully saturated with anaesthetic vapour. Wicks help to increase the surface area of the anaesthetic liquid by drawing it up through capillary action.

Baffles direct the vaporiser flow towards the liquid. This ensures that gas leaving the chamber is fully saturated with anaesthetic vapour and helps maintain accuracy at higher flow rates.

When gas leaves the chamber, energy is lost in the form of latent heat of vaporisation. Because SVP falls with temperature, a reduction in the operating temperature of the vaporiser would affect the concentration delivered to the patient. **Heat sinks** made of copper encase the vaporiser. They have a high specific heat capacity and allow the exchange of thermal energy between the device, vaporising chamber and atmosphere to minimise the temperature disruptions cause by vaporisation.

Applied science

How is the plenum vaporiser output calculated?

The SVPs of anaesthetic agents at room temperature and one atmosphere pressure are far above that required for general anaesthesia. For example, isoflurane has a SVP of 239 mmHg at 20°C, which would yield a gas concentration of (239/760 × 100) = 31.4% in the vaporising chamber. Vaporisers are designed to deliver clinically useful concentrations by altering the splitting ratio. The gas concentration in the vaporising chamber is given by the formula:

$$\textit{Concentration}\ (\%) = \frac{\textit{Saturated Vapour Pressure}\ (\text{mmHg})}{\textit{Ambient Pressure}\ (\text{mmHg})} \times 100\%$$

Newer vaporisers have the control valve located at the vaporising chamber outlet. For these devices the splitting ratio is given by:

$$\textit{Split Ratio} = \frac{100 - \textit{Dialed Concentration}}{\textit{Dialed Concentration}} \times \frac{\textit{SVP Agent}}{\textit{Ambient Pressure}} - 1$$

Using the equation above, if the isoflurane concentration control dial is set to 1%, the splitting ratio would be 30.1 (99 × 0.314 – 1) to 1. With a fresh gas flow of 4000 ml·min^{-1}, 128.6 ml·min^{-1} would constitute the vaporiser flow and the bypass flow would be 3871 ml·min^{-1}. To calculate vaporiser output, the following relationship is used:

$$\frac{\textit{SVP Agent}\ (\text{mmHg})}{\textit{Ambient Pressure}\ (\text{mmHg})} = \frac{\textit{Vapour}\ (\text{ml})}{\textit{Carrier Gas}\ (\text{ml}) + \textit{Vapour}\ (\text{ml})}$$

Rearranging the equation above to solve for vapour will yield the volume of volatile entrained by a specific vaporiser flow. If the vaporiser flow (carrier gas) is 128.6 ml·min^{-1}, it will pick up 58.9 ml·min^{-1} of isoflurane vapour; therefore, the total volume of gas leaving the vaporiser set at 1% will be 4058.9 ml·min^{-1}.

Questions

What different types of vaporiser are there?
What is the splitting ratio?
Why is temperature compensation required?
What is the purpose of wicks and baffles in the vaporising chamber?

Chapter 12

Fluid dynamics: laminar and turbulent flow

Basic science

Flow is the volume of fluid passing a fixed point per unit time. Fluid flow can be either laminar or turbulent.

Laminar flow describes smooth flow, which occurs in parallel layers. For laminar flow in a tube of uniform diameter, the fastest flow velocity takes place in the centre of the stream, while the slowest occurs adjacent to the walls due to frictional forces.

The **Hagen–Poiseuille** law (equation below) is used to describe laminar flow in a tube of uniform diameter.

$$Q = (\Delta P \cdot \pi \cdot r^4)/(8 \cdot \eta \cdot l)$$

where Q is flow, ΔP is pressure gradient, r is radius, η is fluid viscosity and l is length.

From this equation, it can be seen that flow is directly proportional to the pressure gradient and the fourth power of the radius. Therefore, doubling the diameter of the tube will increase flow by 16 times (2^4). Flow is inversely proportional to the viscosity of the fluid and the length of the tube.

Turbulent flow is rough and disordered. Eddy currents and parallel mixing occur within the fluid stream.

The **Reynolds number** is a dimensionless number that is defined as the ratio of inertial and viscous forces. It is used to predict which type of flow is likely to be present. A Reynolds number <2000, where viscous forces predominate, predicts flow to be laminar. Between 2000 and 4000, both laminar and turbulent flow are anticipated. Above 4000, flow is likely to be completely turbulent because inertial forces are dominant. **Critical flow** is the point above which turbulent flow commences, which occurs at a Reynolds number of around 2000.

$$\textit{Reynolds Number}\ (\mathrm{Re}) = \frac{\rho \cdot v \cdot d}{\eta} = \frac{\textit{Inertial Forces}}{\textit{Viscous Forces}}$$

where ρ is fluid density, v is velocity, d is diameter and η is fluid viscosity.

Applied science

How can knowledge of flow be used in the management of upper airway obstruction?

In partial upper airway obstruction, as gases flow through the constriction, flow velocity increases. This increases the likelihood of turbulent flow occurring, which is predicted by an increase in Reynolds number. Because turbulent flow is chaotic and less organised than laminar flow, it is less efficient at delivering gas to the alveoli.

Knowledge of Reynolds number is beneficial when attempting to improve gas delivery in cases of partial upper airway constriction. Manipulation of variables to decrease Reynolds number will increase the likelihood of laminar flow occurring with subsequent improvement of gas delivery to the alveoli–capillary interface.

The viscosities for air, oxygen and helium are similar, so they will have minimal effect on changing flow characteristics. However, it is possible to reduce the density of the inspired mixture to increase the likelihood of laminar flow occurring. Helium is less dense than both air and oxygen. **Heliox** is a mixture containing 79% helium and 21% oxygen, and is 70% less dense than air. When used for airway obstruction it improves flow by up to 1.73 times. However, this does not necessarily improve oxygenation as the fraction of inspired oxygen (FiO_2) is only 0.21. Conversely, nitrous oxide has a higher density than oxygen and should be used with caution in airway obstruction because it could increase turbulence.

In the preoperative patient, increasing the diameter of the constriction may not be possible without surgical intervention, unless it was due to oedema. Finally, a reduction in flow velocity could also decrease Reynolds number, but as this would reduce overall flow it would be counter-productive.

Questions

What is flow?

Explain the difference between laminar and turbulent flow.

What is the Hagen–Poiseuille equation?

What factors affect flow when it is turbulent?

What is Reynolds number? What is critical flow?

Chapter 13

Fluid dynamics: Bernoulli and Venturi

Basic science

The **Bernoulli principle** states that as an ideal fluid flows through a constriction, there is an increase in flow velocity and a decrease in the pressure exerted by the fluid. It assumes that flow is steady, streamline and that there is no friction.

When fluid flows through a uniform tube under a constant driving pressure, a fixed volume will pass any given point in the tube per unit time. If a constriction is applied in the same tube, there will be an increase in flow velocity. This occurs because the same volume must pass through the narrower segment of the tube per unit time as that passing through the wider section. Therefore, flow velocity varies inversely with the cross-sectional area of the tube.

The **law of conservation of energy** states that energy cannot be created or destroyed, but can be transformed. As fluid flows through a constriction, the kinetic energy of the fluid increases. Assuming the energy loss due to friction is negligible, and that no energy is added or removed from the system, the increase in kinetic energy must come at the expense of potential energy. In other words, as the speed of fluid flow increases, the pressure it exerts falls.

Using the above example, if after the constriction the tube diameter gradually increases to produce an exit cone, the flow velocity at this point would begin to fall and the pressure would increase. An opening at the constriction within the tube would result in entrainment of fluid into the flow stream. This occurs because the pressure at the constriction is less than atmospheric pressure (Bernoulli principle). Entraining fluid via a side port allows flow velocity to be maintained at a higher rate in the exit cone. This is the **Venturi effect**.

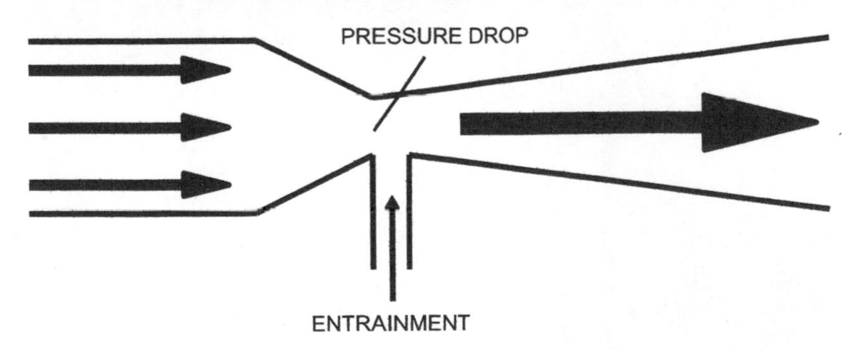

By altering the driving flow, the constriction size or the size of the opening, the Venturi effect can be used to mix gases or fluids to produce a fixed concentration. The **entrainment ratio** is the ratio between the entrained fluid flow and the driving fluid flow.

Applied science

Describe how high-airflow oxygen enrichment masks work

High-airflow oxygen enrichment (HAFOE) masks are also known as fixed performance or venturi masks. Their function is based on the Venturi effect.

These devices use air entrainment to deliver fixed concentrations of oxygen to patients at flow rates that exceed the patient's peak inspiratory flow rate. Oxygen flows through a constriction to create a pressure drop in accordance with the Bernoulli principle. The fall in pressure is used to entrain ambient air through side openings in the venturi device.

The oxygen flow rate required is specified by the manufacturer for different aperture sizes, which result in a predictable fraction of inspired oxygen (FiO_2). They are capable of achieving inspired oxygen concentrations of 24%, 28%, 31%, 35%, 40% and 60% depending on the aperture size and the driving oxygen flow rate.

Flow rates higher than the patient's peak inspiratory flow rate are required to ensure accurate oxygen concentration delivery. The entrainment ratio refers to the amount of air entrained for the driving oxygen flow. A high entrainment ratio produces a relatively low FiO_2, while a low ratio produces a higher FiO_2. To achieve high oxygen concentrations, the entrained volume of air needs to be limited, resulting in a reduced total flow of gas to the patient. Therefore, HAFOE devices are not suitable for the administration of oxygen concentrations greater than 60%, as the gas flow may be inadequate.

Questions

What is the Bernouilli principle?

Why does flow velocity increase and pressure fall at a constriction in a tube?

What is meant by the Venturi effect?

How is this used clinically?

What is the entrainment ratio?

Chapter 14

Diffusion and osmosis

Basic science

Diffusion is the spontaneous movement of a substance from an area of high concentration to an area of low concentration.

The rate of diffusion across a membrane is directly proportional to the concentration gradient (**Fick's law**), membrane area and temperature. It is inversely proportional to the square root of the molecular mass of the fluid (**Graham's law**) and the membrane thickness. Diffusion across cell membranes is also influenced by membrane-bound carrier proteins and ionic charges.

Osmosis is the spontaneous movement of **solvent** across a semi-permeable membrane from an area of low solute concentration to an area of high solute concentration.

Large impermeable molecules on one side of a semi-permeable membrane exert an **osmotic pressure**. The osmotic pressure is the hydrostatic pressure required to prevent osmosis occurring. Osmosis acts to balance the concentration gradients of these large, impermeable molecules.

Osmolarity is the number of osmoles of solute per **litre of solution**. **Osmolality** is the number of osmoles of solute per **kilogram of solvent**. Both osmolarity and osmolality measure the number of osmotically active particles on one side of a semi-permeable membrane. Osmolarity tends to be lower than osmolality because the former is expressed in terms of a volume, which can be affected by the presence of other molecules such as plasma proteins. It is also altered by changes in temperature.

Colligative properties are used to describe how the properties of a solution change with different ratios of solvent to solute. When the osmolality of a solution is increased, the boiling point is elevated, the freezing point is depressed, the saturated vapour pressure falls (**Raoult's law**) and osmotic pressure rises. The fall in vapour pressure and subsequent rise in boiling point occur because the non-volatile solute added occupies some of the space at the surface of the solution. This prevents the solute molecules from entering the vapour phase as easily as they did before. Freezing point depression occurs because the solute molecules do not form a tight lattice structure as readily. Therefore, lower temperatures are required to cause freezing.

Applied science

How do we measure the osmolality of solutions?

When the osmolality of a solution changes, the colligative properties of the solution also change in a predictable manner. An **osmometer** can measure the changes in colligative properties and relate them to changes in osmolality. The most commonly used osmometers measure the depression of freezing point caused by solute in a solution. The osmolality of a solute containing solution can then be calculated by comparing the measured freezing point against that of its pure, solute-free counterpart.

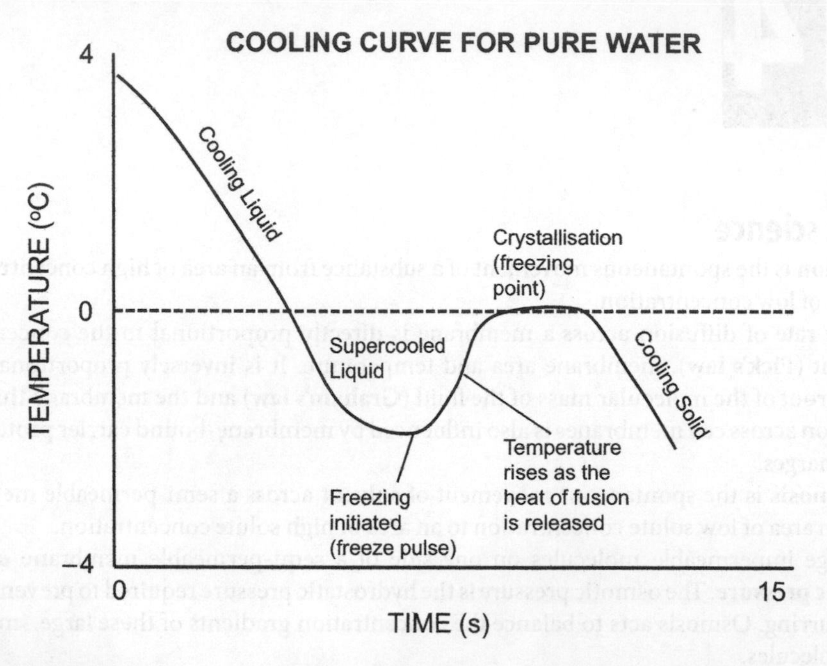

The osmometer uses **supercooling** to measure the depression of the freezing point. Supercooling is the process of cooling a liquid without causing a change in state. It involves rapidly cooling a sample to a temperature below its expected freezing point. The substance does not form a solid structure because freezing requires a seed crystal or nucleus around which a crystal structure can form. Once the desired temperature is reached, a physical shock called the 'freeze pulse' is introduced, which helps initiate rapid freezing by aligning some molecules to create a crystal seed. The crystallisation process releases the heat of fusion, which causes the temperature of the sample to rise. The temperature increases until equilibrium between the solid and liquid phases is established. At this point, the temperature represents the freezing point.

Questions

What is diffusion?

What is osmolality and osmolarity? Why are they different in plasma?

What is supercooling?

Chapter 15

Surface tension

Basic science

Neighbouring molecules within a liquid are subject to intermolecular forces of attraction. In the bulk of a fluid, molecules are exposed to a greater number of **cohesive forces** as compared to those at the **surface interface** (the area where different fluids meet).

Surface molecules are in an energetically unfavourable state; they have more free energy than those in the bulk of the fluid. The addition of energy by thermal agitation overcomes the intermolecular forces of attraction, causing those molecules at the surface to pass into the gaseous phase. The creation of a new surface requires energy. Therefore, a fluid system will naturally adopt a shape that has the smallest surface area possible.

Surface tension occurs at the interface between fluids; for example, air and water. Relatively higher forces of attraction at the surface interface exist because the outermost layer of molecules has an excess of free surface energy. The molecules in the liquid have a greater attraction to neighbouring molecules (cohesion) than to the air molecules (adhesion), so they are pulled in towards the bulk of the fluids as well as towards the molecules at each side. As a result, the liquid molecules form to fill the least surface area possible, and the surface layer acts like a film covering the surface of the fluid. For this reason a drop of water adopts the shape of a sphere.

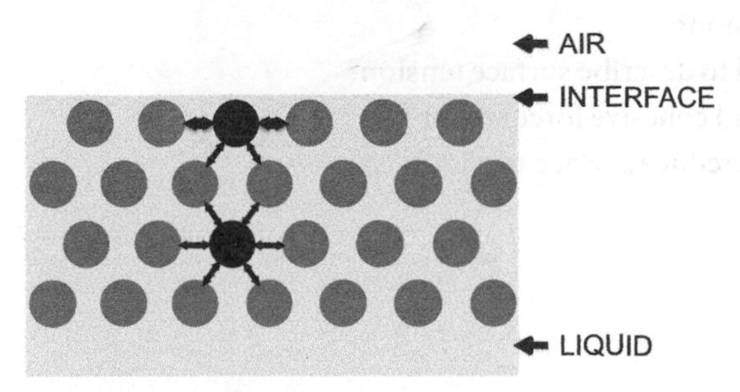

Surface tension is commonly measured as **force per unit length** or **energy per unit area.** The curved meniscus formed in column of mercury in the liquid barometer is an example of surface tension.

Applied science

Describe Laplace's law and give an example of its application

In medicine, the relationship describing pressure, wall tension and radius of cylindrical shapes and spheres is known as **Laplace's law**.

For spherical structures, e.g. alveoli and aneurysmal vessel segments:

$$Wall\ Tension = \frac{Transmural\ Pressure \times Radius}{2}$$

For cylindrical structures, e.g. blood vessels:

$$Wall\ Tension = Transmural\ Pressure \times Radius$$

Surfactants are used to reduce surface tension. They are amphiphilic molecules with hydrophilic heads and hydrophobic tails. Lining the alveoli, there is a thin film of water that has a high surface tension and tends to collapse the alveoli inwards. Pulmonary surfactant prevents alveoli from completely collapsing and improves their compliance. This occurs because the hydrophilic heads of the surfactant **adsorb** into the water surface interface, disrupting the effect of surface tension. The hydrophobic tails keep surfactant at the interface by preventing it from dissolving into the water.

It can be seen from Laplace's law that an alveolus with a large radius needs a lower pressure to overcome wall tension to cause expansion. Application of a positive pressure will likely cause expanded alveoli to distend further than completely collapsed alveoli. It is similar to inflating a balloon; initially, high pressures are required to commence inflation, but as it expands the pressure required to inflate it further falls. This is why patients with pulmonary bullae are prone to developing pneumothoraces with high positive peak inspiratory pressures.

Questions

What is surface tension?
What units are used to describe surface tension?
How do adhesive and cohesive forces differ?
How do surfactants reduce surface tension?

Chapter

16 Medical gases: storage and supply

Basic science

Medical gases are supplied at high pressure via cylinders or pipeline from a cylinder manifold, a vacuum-insulated evaporator (VIE) or gas concentrators.

Medical gas cylinders contain compressed gas at high pressure. Most cylinders are made of **molybdenum steel** (steel plus an alloy). Cylinders manufactured from aluminium are also available and can be safely used in an MRI scanner.

Correct identification of a cylinder and its content are vital to prevent adverse incidents associated with use. Important features of medical gas cylinders include the following.

- Colour coding of the neck of the cylinder can assist in identification of the cylinder, but should never be used in isolation. The cylinder label must be used to correctly identify content. The British Compressed Gases Association (BCGA) members will supply all medical gases in coloured-neck and white-bodied cylinders by 2025 to comply with internationally agreed standards.
- Labels placed on the shoulders of cylinders contain vital information, such as hazards and precautions, hazard diamonds, product name, filled pressure, gross weight, cylinder size and serial number.

Pin-index safety systems and gas-specific connectors are used on small and large cylinders (sizes G and H), respectively, to prevent the incorrect attachment of gas yolks and regulators. Pins fitted to a gas yolk correspond to holes in the cylinder value. The positions are gas-specific, so that only correct pairing will result in tight seal.

GAS	PIN INDEX
OXYGEN	2-5
NITROUS OXIDE	3-5
AIR	1-5
CARBON DIOXIDE	1-6
HELIOX	2-4

Applied science

How do you determine how much gas remains in a cylinder?

For cylinders containing compressed (non-liquefied) gas, gauge pressure falls as gas is used; Boyle's law can be used to calculate how much gas a cylinder will deliver at ambient pressure. Oxygen cylinders are an example.

Liquefied gas cylinders such as nitrous oxide and carbon dioxide contain liquid and vapour. In this instance, the gauge pressure only displays the vapour pressure, which does not relate to the amount of substance remaining in the cylinder. To determine how much remains, weight is used. **Tare weight** is the weight of the empty cylinder. The difference between the filled cylinder weight and tare weight indicates the amount of substance present. Taking nitrous oxide as an example, its molecular weight is 44 g. According to Avagadro's hypothesis, one mole of nitrous oxide weighs 44 g and occupies 22.4 l at STP. A cylinder containing 2000 g of nitrous oxide will therefore contain 1018 l ([2000 g/44 g] × 22.4 l).

The gauge pressure of a nitrous oxide cylinder normally reads 4400 kPa (the SVP of nitrous oxide at 20°C). During use, vaporisation of liquid requires energy for the latent heat of vaporisation. This causes the temperature of the remaining liquid to fall, resulting in a decrease in vapour pressure within the cylinder. The gauge pressure falls to reflect this, which can be misinterpreted as a cylinder that no longer contains liquid. However, when cylinder use is ceased, temperature and vapour pressure increase to their original value.

The **filling ratio** is the weight of fluid in the cylinder divided by the weight of the water required to fill it. Liquefied gas within cylinders will exert a greater pressure when warm. To avoid cylinder pressures exceeding safety limits, cylinders are not completely filled, having a filling ratio of 0.75 in the UK and 0.67 in hotter climates.

Questions

What are gas cylinders made of?
What information is provided on cylinder labels?
What is meant by the term tare weight?
What is the filling ratio for nitrous oxide?
What is the pin-index safety system?

Chapter 17

Electricity – basics

Basic science

Electricity is a form of energy resulting from the presence of charged particles. It can be dynamic, as the flow of charge particles, or static, by the accumulation of charge.

Magnetism is the term used to refer to the attractive and repulsive forces between objects caused by the movement of electric charge.

An electric current is the flow of electrically charged particles (electrons) through a substance. These substances can be classified as conductors, semi-conductors or insulators depending on their ability to allow a charge to move through them.

Conductors allow electrical charge to move freely through their structure. The outer electrons of the atoms are loosely bound and free to move through the material in response to an electric potential difference. Examples include metals such as iron. **Insulators** do not allow electrical charge to move freely because their electrons are more firmly bound. Examples of insulators include rubber, glass and wood. **Semi-conductors** have conductivity between that of an insulator and conductor. Semi-conductors are used widely in electrical components such as solar cells, transistors, thermistors and light-emitting diodes (LEDs).

An electric circuit is a closed path of conductors (e.g. copper wires) and electrical components in which electric current flows under the influence of a voltage or current source. Circuits can be in series or parallel. In a **series circuit**, there is only one route for the flow of electrons between the negative and positive terminals. In a **parallel circuit** there are multiple routes through which electrons can flow. The proportion of electrons taking each route will depend on the resistance of each route to the flow of current.

The **electrical charge** of matter can be either positive or negative and is related to the number of excess protons or electrons, respectively. The SI unit for electrical charge is the **coulomb** (C). Each electron carries a negative charge of $1.60217657 \times 10^{-19}$ C and one coulomb can be considered to have a charge of roughly 6.241×10^{18} electrons. A material that loses 6.241×10^{18} electrons will have a positive charge of one coulomb, whereas a material that gains 6.241×10^{18} electrons will have a negative charge of one coulomb.

Electrical power is measured in watts (SI unit) and is equivalent to one joule of energy transferred in one second. In terms of electricity, one watt is the rate at which work is done when one ampere of current flows through an electrical potential difference of one volt. Simply, electrical power is the rate of energy use, measured in watts.

$$Power = \frac{Energy}{Time}$$

$$1\ watt\ (W) = \frac{1\ joule\ (J)}{1\ second\ (S)} = 1\ volt\ (V) \times 1\ ampere\ (A)$$

An **electric current** is formed by the flow of charge carriers through a conducting material, similar to water flowing through a pipe. The base SI unit ampere (A) is defined in terms of the current flow required to generate a specific force of 2×10^{-7} newtons between two wires of infinite length one metre apart in a vacuum. When current flows in a wire, an electromagnetic field is produced. As adjacent magnetic fields interact, the two wires with current flowing will exert an attractive or repulsive force upon each other. One ampere of current represents one coulomb of electrical charge (6.241×10^{18} charge carriers) moving past a cross-section of conductor in one second.

$$Electric\ Current = \frac{Electric\ Charge}{Time}$$

$$1\ ampere\ (A) = \frac{1\ coulomb\ (C)}{1\ second\ (S)}$$

Voltage is measured in volts (SI unit). It is the 'driving force' or 'pressure' which causes electrons to flow. The volt is defined as the potential difference between two points in an electric circuit that will impart one joule of energy per coulomb of charge that passes through it.

$$Voltage = \frac{Energy}{Electric\ Charge}$$

$$1\ volt\ (V) = \frac{1\ joule\ (J)}{1\ coulomb\ (C)}$$

Applied science

What is resistance? What are resistors?

Resistance is the characteristic of a material that impedes the flow of electrons. All materials have some degree of resistance; the amount of resistance depends on the individual properties of the material. The flow of electrons through a resistant material generates heat.

Ohm is the unit of measure for electrical resistance. It is defined as the resistance between two points when a constant potential difference of one volt produces a current of one ampere. This relationship is known as Ohm's law. Mathematically, the relationship between electrical current, resistance and voltage is described as:

$$Resistance\ (ohm) = \frac{Potential\ Difference\ (volt)}{Current\ (ampere)}$$

Resistors are components that add resistance to electrical circuits to reduce current flow. The resistance of a resistor is the same for alternating current (AC) and direct current (DC), because its resistance is independent of the frequency of the current through it.

Resistors can be designed and manufactured with different operating properties to obtain a desired function. Resistors can have a nearly constant resistance over an operating temperature range, or can have a gradient with temperature. The latter are called **thermistors** and find use in temperature control circuits. **Voltage-controlled resistors** (**VCR**) have a resistance proportional to the voltage across them.

Joule's first law describes the relation between heat energy (Q) generated by an electric current (I) passing through a conductor of resistance (R) for an amount of time (t). This property of resistors forms the basis of electric heaters.

$$Q = I^2 \cdot R \cdot t$$

What is impedance? How does it differ from resistance?

Impedance and resistance are terms used to describe the opposition to electrical current flow. Resistance is used to describe opposition to flow in DC circuits, whereas impedance is used for AC circuits. Unlike direct currents, alternating currents exhibit reactance (capacitive and inductive) because they have frequency and phase associated. In AC circuits, **impedance** (Z) consists of a real part (ohmic resistance) and an imaginary part (reactance of an inductor or capacitor). It is measured in ohms, and is the total opposition to current flow. Mathematically, it can be expressed as a complex number:

$$Impedance\ (Z) = Resistance\ (R) + [\sqrt{-1}\ (j) \times Reactance\ (X)]$$
$$Z\ (modulus\ or\ magnitude) = \sqrt{R^2 + X^2}$$

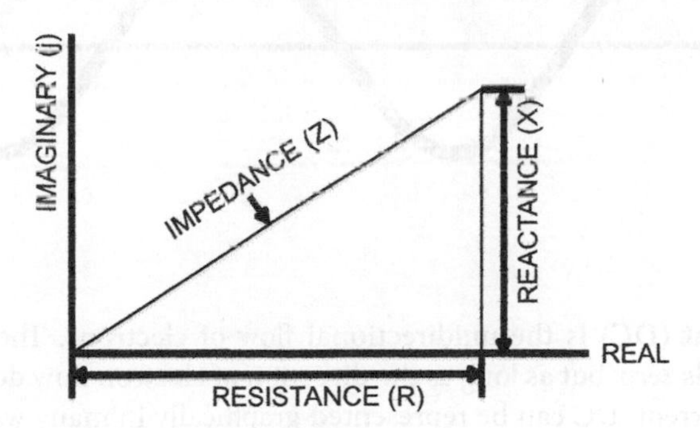

The magnitude of impedance, for a given resistance and reactance, is equal to the square root of the sum of the squares of the resistance and the reactance, expressed in ohms (as per Pythagorus' theorem).

Questions

Describe the terms electricity and magnetism.
What is the difference between conductors and insulators?
What are semi-conductors used for?
Describe the difference between series and parallel circuits.
What is meant by the terms voltage, power, charge, current?
Give some equations for each of the above.
What is Ohm's law?

Chapter 18

Electricity – AC and DC

Basic science

An electrical current is the flow of electrons. The direction of flow determines whether the current is an alternating or direct current.

An **alternating current** (AC) occurs when the flow of electrons in a conductor or semi-conductor periodically reverses direction. The frequency (hertz) of AC is the number of complete cycles per second. A sine wave can be used to illustrate an alternating current.

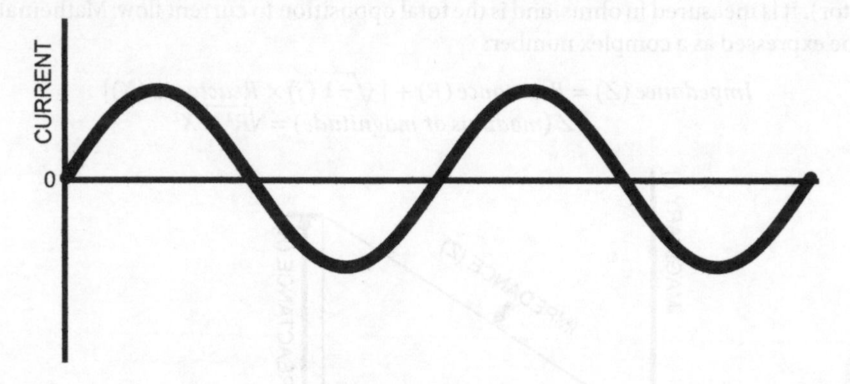

A **direct current** (DC) is the unidirectional flow of electrons. The current flow may vary and fall towards zero, but as long as the direction of electron flow does not change, this remains a direct current. DC can be represented graphically in many ways; one example is shown below.

AC can be converted to DC by a rectifier circuit (shown below). **Diodes** are used in rectifier circuits because they conduct one way (behaving as one-way flow valves for the electric current). This property of the diode is the fundamental principle of designing DC power supplies from AC.

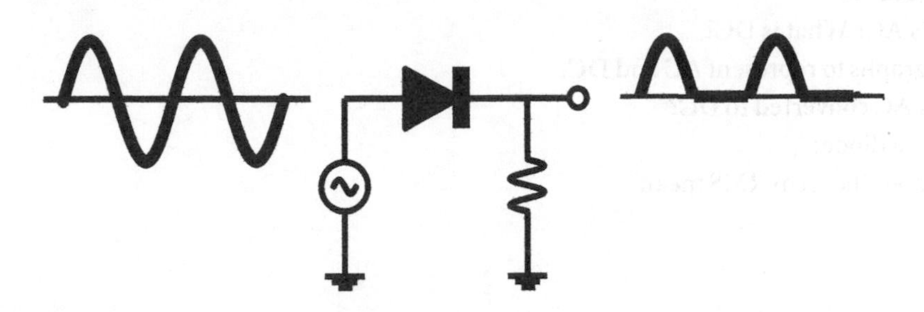

Applied science

How can we compare AC to DC?

Measuring the voltage of DC can be done directly, but measuring the average voltage of AC is not as simple. Because the waveform is sinusoidal, AC does not have a steady voltage. It has equal deflections on both the positive and the negative axis, so the mean voltage is zero. The magnitude of an AC can be expressed as peak voltage or a **root mean squared (RMS)** value. In the UK, the domestic mains supply is 240 V RMS, but the peak voltage is actually 339 V ($240\sqrt{2}$).

The **root mean squared (RMS)** value allows comparison of the equivalent amounts of AC and DC power required to produce the same heating effect.

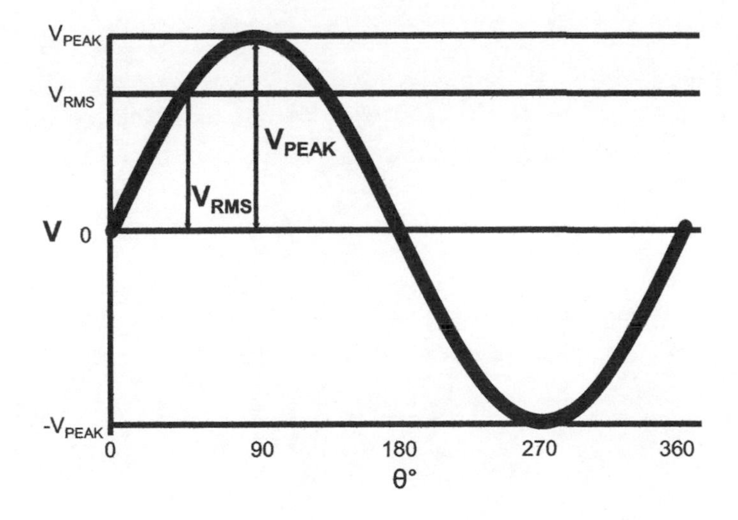

A sinusodial waveform is represented by $V_{PEAK} \cdot \sin\theta$. To calculate the RMS, the square root of the mean of the squares of all the amplitudes for θ from 0° to 360° is calculated. The mean of $\sin^2\theta$ from 0° to 360° equals 1/2. Therefore, RMS is given by $V_{PEAK}/\sqrt{2}$.

Questions

What is AC? What is DC?

Draw graphs to represent AC and DC.

How is AC converted to DC?

What is a diode?

What does the term RMS mean?

Chapter 19

Electricity – generation and distribution

Basic science

Electricity for UK mains distribution is produced by alternators (electrical generators) in power stations and distributed by a system of conductors. An **alternator** consists of a rotating magnetic core surrounded by stationary windings of a conductor (coils of wire). Rotation of the magnet results in a fluctuating magnetic field, which induces an **electromotive force** in the surrounding windings. If a circuit exists, this electromotive force will move the charge (electrons) present in the wires. The fluctuating nature of the magnetic field results in a current that varies sinusoidally (alternating current). In the power station, three-phase alternators are used, which have three sets of windings (120° apart) surrounding a rotating magnetic core. These alternators will produce three sinusoidal waveforms which are out of phase with one another by the time taken for the magnet to rotate 120°. This is known as **three-phase AC**.

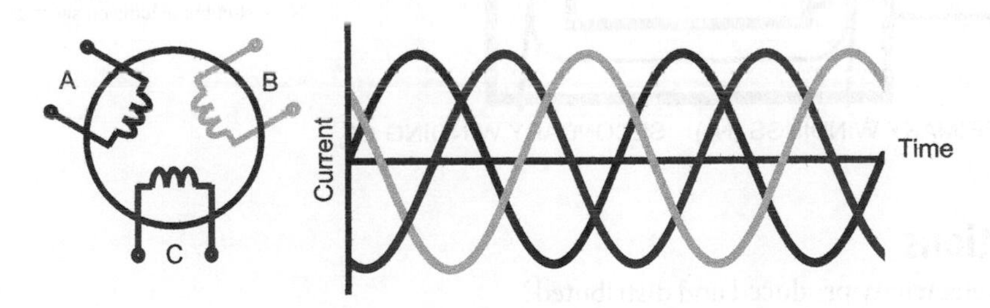

Three-phase power leaves the power station via three wires (one for each phase of AC) to enter a step-up substation. Here, **transformers** convert the power station output to a very high voltage (155 000–765 000 V) for **transmission** in high-voltage power lines over long distances. High voltages are used for transmission and distribution to maintain efficiency, because power losses in the grid are related to current flow, not voltage. As electricity approaches the end user it passes a series of step-down substations, where the voltage is decreased to around 10 000 volts for **distribution** to substations close to the users. Here, the voltage is decreased to 240 V RMS in each of the three phases. At this substation, one limb from each phase leaving the transformer join at the **star point** to form a neutral and ground common to all three. The other end of the circuits for each phase are used to supply the 'live' to properties within range of the substation. The live entering a property is only of a single phase.

Applied science

What are the transformers? Describe their principle of operation

Faraday's law describes how alterations in a magnetic field around any closed circuit will generate an electromotive force (potential difference) to cause current flow. According to Faraday's law, the greater the change in the magnetic field over time, the greater the electromotive force. This principle is fundamental to the operation of transformers.

Transformers transfer electrical energy between two circuits by electromagnetic induction. They consist of two sets of windings (primary and secondary) around a ferrous core. Alternating current flow in the primary windings induces a fluctuating magnetic field, which in turn creates a fluctuating magnetic flux in the core. This induces an electromotive force in the secondary windings, causing current to flow. The ratio of the number of coils between the primary and secondary windings determines the output voltage.

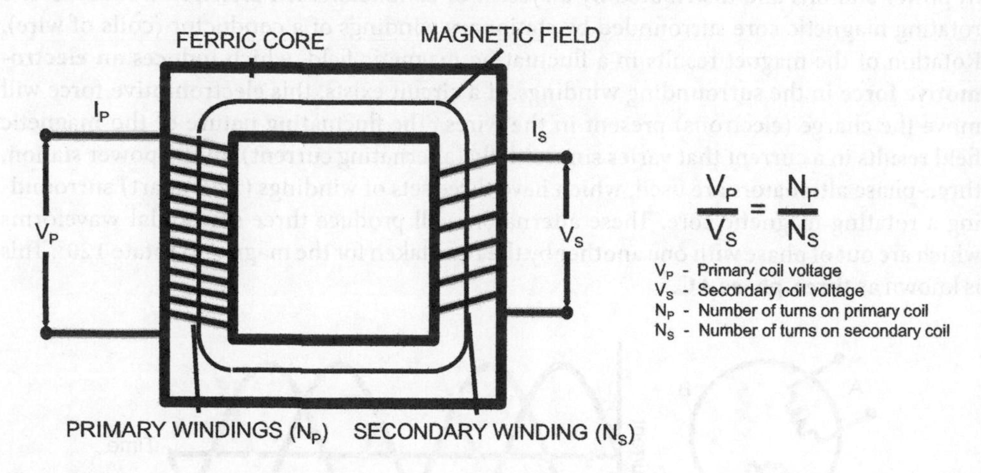

Questions

How is electricity produced and distributed?
Why are high voltages used for transmission?
What occurs at the substations?
What is meant by the term three-phase AC?
What is the star point?
What is Faraday's law?

Chapter 20

Electrical safety

Basic science

Hazards of electricity include **shock** (microshock and macroshock), **thermal risks** (burns, fires and explosions) and **electrical interference**.

Electrical current is transmitted through the body by the electrons associated with free ions in the extra- and intracellular spaces. DC and low-frequency AC cause movement of ions across cell membranes, which disrupt normal ionic gradients and cell function. Myocytes and neurones are most susceptible to electrical current, due to their relative abundance of ion channels and pumps. The movement of ions and electrons also generates heat, which has the potential to cause thermal injury.

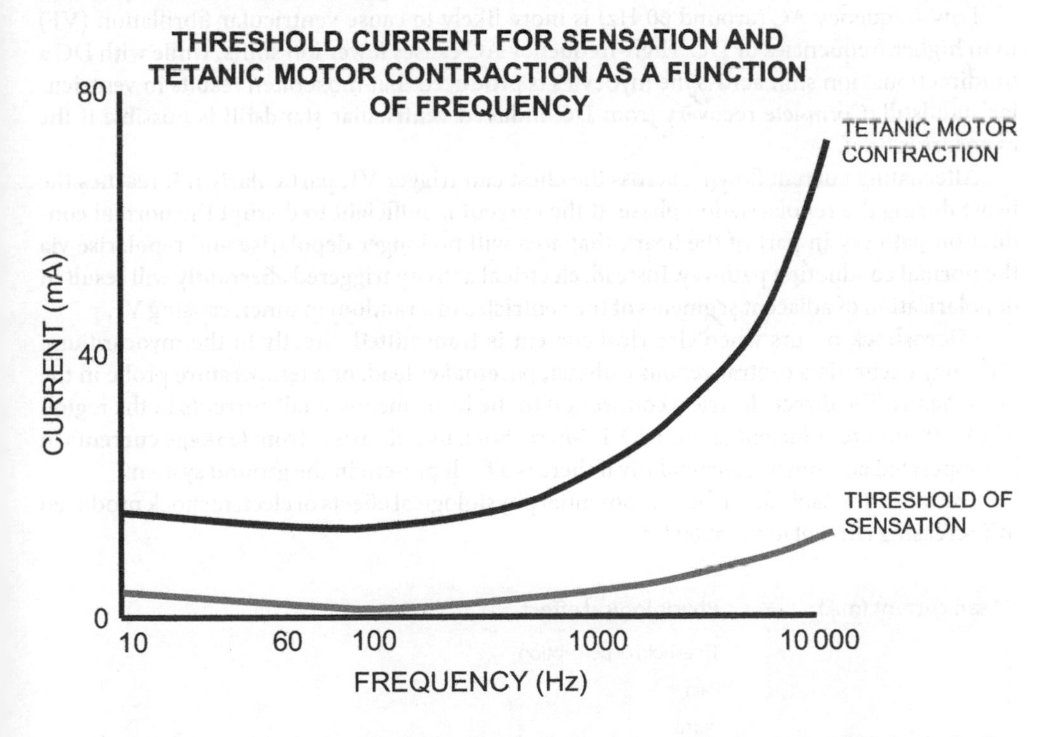

The graph above depicts the relationship between current, frequency and biological effect on sensation and muscle contraction. At high frequencies, greater current flow is required to cause muscle contraction and sensation than at 60–100 Hz. This is because at high frequencies, polarity reverses too rapidly to allow time for ion movement. This principle explains the relative safety of diathermy, as it uses high-frequency current to minimise the electric shock risk. However, high-frequency AC is still capable of producing significant heating effects leading to thermal injuries.

Shock hazards can occur via resistive or capacitative coupling. **Resistive coupling** is the process where an earthed individual completes a circuit, by making direct contact with a live object, such as the casing of faulty equipment. **Capacitative coupling** occurs when an object is close to a strong AC source, such as an operating light. The AC source and the object form two plates of a capacitor, separated by air (the dielectric). This allows alternating current to flow, resulting in shock and thermal hazards.

Macroshock occurs when a person forms part of an electrical circuit, resulting in current flow through them. This can cause arrhythmias, involuntary muscle contraction (including respiratory muscles), and altered nervous system function (confusion and coma). The severity of macroshock depends on the strength and type of current, the path of current flow through the body, the duration of exposure and the timing of shock in relation to the cardiac cycle.

Current strength is determined by potential difference divided by impedance. Skin and clothing are the main sources of impedance. For a given potential difference, the amount of current flow will be greater if the impedance is low rather than high. Wet skin has lower impedance than dry skin, because water usually contains free ions within it, which improves conduction. Increased contact area and intravenous lines (breaching the skin) also lower impedance.

Low-frequency AC (around 60 Hz) is more likely to cause ventricular fibrillation (VF) than higher frequencies or DC. High-frequency AC causes fewer ion shifts, while with DC a unidirectional ion shift across the myocytes is produced that most often results in ventricular standstill. Complete recovery from DC-induced ventricular standstill is possible if the exposure is brief.

Alternating current flowing across the chest can trigger VF, particularly if it reaches the heart during the repolarisation phase. If the current is sufficient to disrupt the normal conduction pathway in part of the heart, that area will no longer depolarise and repolarise via the normal conduction pathway. Instead, electrical activity triggered aberrantly will result in depolarisation of adjacent segments of the ventricles in a random manner, causing VF.

Microshock occurs when electrical current is transmitted directly to the myocardium. This may occur via a central venous catheter, pacemaker lead, or a temperature probe in the oesophagus. The direct electrical connection to the heart means small currents in the region of 50–100 μA are sufficient to induce VF. Microshock usually arises from **leakage currents** in line-operated equipment, particularly if there is a fault present in the ground system.

The following table describes the potential physiological effects of electric shock produced by increasing current levels at 60 Hz.

Mean current (mA)	Physiological effect
1	Threshold of perception
5	Pain
8	Burns
15	Muscle contraction
50	Respiratory arrest and severe burns
100	Ventricular fibrillation
0.05 (50 μA)	Microshock (if cardiac connection exists)

Thermal hazards associated with electricity include burns, fires and explosions.

Burns occur when a current passes through a small area of skin, creating a **high current density**. This causes heating, and is the principle underpinning diathermy. Burns are more likely with high-frequency AC (over 100 kHz) and high current flow.

Fires and explosions occur when electrical sparks ignite flammable liquids or combustible gases, such as alcohol-based cleaning solutions and oxygen. **Static electricity** is the accumulation of electric charge on an insulated surface, which occurs when electrons are transferred from one object to another by direct contact, for example rubbing materials together. A spark may form if the charged object is placed near to a conductor such as an earthed object. This occurs because the charge is able to ionise surrounding air molecules in order to conduct across the air gap. Lightning strikes are an example of electrostatic discharge between areas of charged cloud and the Earth's surface.

Applied science

What measures are used to reduce the risk of shock in theatre?

Measures to reduce risks posed by the **theatre environment** include cleaning **saline spills**, which reduce impedance to current flow, ensuring patients are not in contact with earthed objects, and maintaining **humidity** at >50% to prevent static accumulation.

Equipment selection and annual maintenance to detect and repair faults reduces risk of shock. Theatre tables are non-conducting, and anti-static shoes have a resistance of between 75 kΩ and 10 MΩ, which prevents accumulation of static charge by allowing conduction of small charges. **Closed breathing systems** and **scavenging** help to keep combustible gases away from sparks, and **carbon-filled rubber** in breathing circuits conducts static charges.

Floating circuits can provide additional safety by electrically isolating devices from the mains supply. Transformers are used in floating circuits to transfer the electrical energy between the two circuits by electromagnetic induction. Given that the isolated circuit is physically separated from the substation electricity supply, current from a device fault will not flow back via the ground/earth system. This is because a complete circuit is not formed. Shock hazards are still possible, but this requires an individual to complete the circuit by making contact with the live and the return line from the isolating transformer.

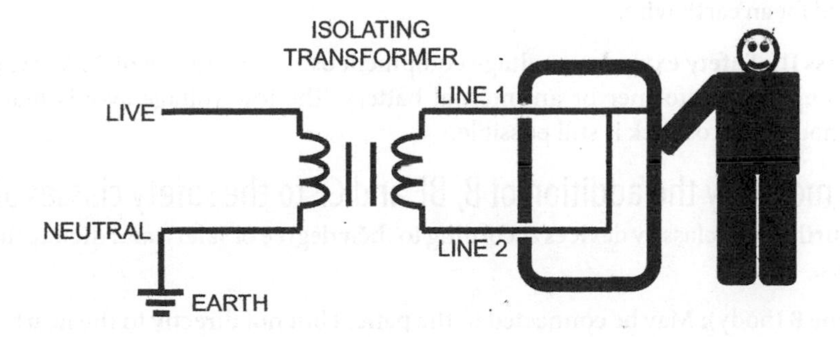

The diagram above depicts a device fault in a floating circuit, where the casing has become live. A person in contact with line 1 (the live case) will not be electrocuted, even if they are earthed, as the current has no path to return to line 2. If the person simultaneously contacts lines 1 and 2, they complete the circuit, receiving a shock.

How are medical electrical devices classified?

The International Electrotechnical Commission (IEC) 60601 describes the safety standard classifications for medical electrical equipment:

Class I: Device casing is **earthed** or **grounded** via the mains plug. If a fault causes the casing to become live, most of the current will preferentially flow to earth via the earth wire, rather than through a person touching the device. This is because the earth provides a lower resistance path for current to flow back to the substation. The resulting current surge melts the **fuses** present on the live, neutral and mains plug, breaking the circuit. **Circuit breakers** can be used to interrupt faulty circuits by detecting excess current flow in fault conditions and disconnecting the supply. Unlike fuses, which require replacement, they can be reset and re-used.

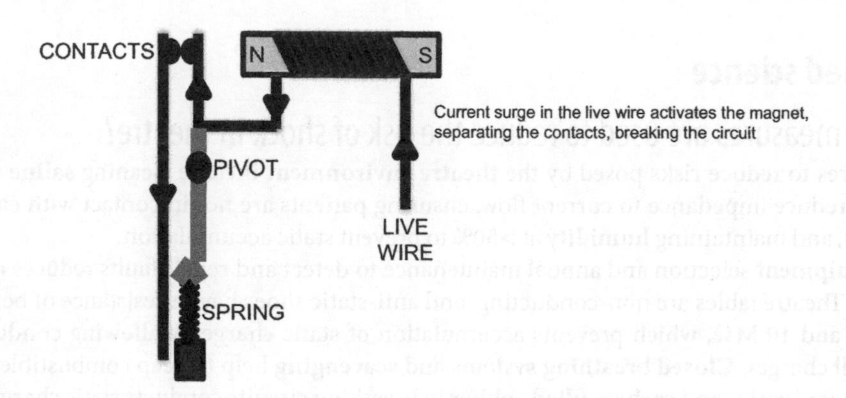

A person touching a faulty earthed device may still be electrocuted if they are in contact with another earthed object with lower impedance, as a greater proportion of current flows via the lower resistance path.

Equipotentiality and **electrical bonding** reduce this risk by maintaining electrical devices and conducting surfaces (e.g. exposed water pipes) at the same potential. This is achieved by linking them to a common earth connection at the mains.

Class II: Double insulation prevents casing becoming live in a single fault, negating the need for an earth wire.

Class III: Safety extra-low-voltage equipment uses a maximum of 24 V AC provided through a transformer or an internal battery. The low voltage avoids macroshock, although microshock is still possible.

What is meant by the addition of B, BF and CF to the safety classes above?

The IEC further sub-classify devices according to their degree of safety, and dictate their usage limitations:

Type B (body): May be connected to the patient but not directly to the heart.

Type BF (body floating): Similar use to type B equipment, but incorporates a **floating circuit**, so is safer.

 **Type CF** (cardiac floating): Safe for connection to the heart and must incorporate a floating circuit.

Leakage currents are possible even in the absence of a fault, because no insulation is perfect. These currents can result in macro- and microshock if they flow through a patient. The IEC further define the **maximum allowable leakage currents** for medical devices during normal conditions and single fault conditions. These are summarised in the table below.

	Type B and BF		Type CF	
	Normal (mA)	Single fault (mA)	Normal (mA)	Single fault (mA)
Earth	0.5	1.0	0.5	1.0
Enclosure	0.1	0.5	0.1	0.5
Patient	0.1	0.5	0.01	0.05

Earth: current flow through the earth wire to ground.

Enclosure: current from an exposed conductive part to earth (not via the dedicated earth wire but another route).

Patient: current flow through a patient connected to the device.

Questions

What influences the damage caused by electrical shock?

What are resistive and capacitative coupling?

Why are some electrical devices earthed?

What is a floating circuit?

Chapter 21 Capacitance, inductance and defibrillators

Basic science

Electric charge is measured in **coulombs**, where one coulomb is the amount of charge passing a point when one ampere flows for one second. One coulomb has a charge of 6.24×10^{18} electrons.

Capacitors are passive components that store electrical energy. The ability of the capacitor to store charge is its **capacitance** and is measured in **farads** (F). The capacitor consists of two terminal plates separated by a non-conducting substance (dielectric). The electrical charge on either plate is proportional to the voltage across the plates of the capacitor. The charge a capacitor can store is related to the dielectric material used **(dielectric constant)**, proportional to the size of the plates and inversely proportional to the distance between them. The way a capacitor behaves electrically is defined by the following relationship:

$$\textit{Charge}\ (\mathrm{C}) = \textit{Capacitance}\ (\mathrm{F}) \times \textit{Voltage}\ (\mathrm{V})$$

Capacitors placed in DC circuits create an increasing resistance to current flow. This is because the charge at the negative plate of the capacitor accumulates, until maximum capacitance is reached and current flow ceases. Capacitors in AC circuits allow current to flow, as the alternating direction of current flow prohibits a significant build up of charge on one of the plates. **Reactance** is the resistance to AC that a capacitor or inductor exhibits and is inversely proportional to frequency. This principle is used in filters to screen out DC currents and low frequency AC.

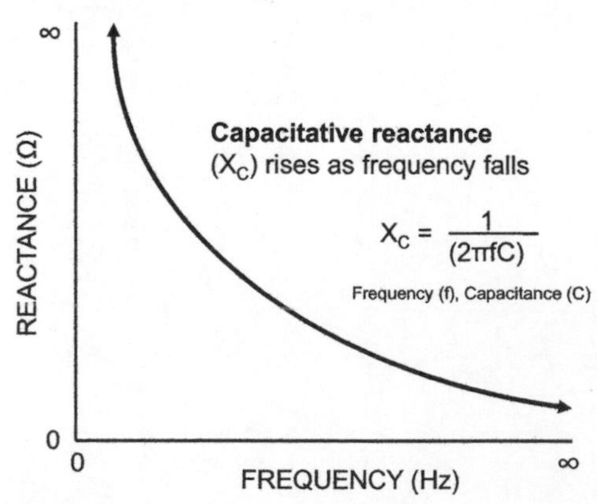

Inductors are coils of conducting wire wound around a ferrous or air core. An increasing current flowing through an inductor generates a magnetic field around it. This magnetic field in turn creates an electromagnetic force, which opposes the current flow, known as **back-emf**. This effect is known as **inductance**, and its SI unit is the **henry (H)**. In a circuit where the rate of current change is 1 A/s, an inductance of one henry would generate one volt across the inductor.

$$Henry\ (\mathrm{H}) = \frac{Voltage\ (\mathrm{V}) \times Time\ (\mathrm{s})}{Amperes\ (\mathrm{A})}$$

When the power is switched off, collapse of the magnetic field induces flow of electrons in the inductor and circuit, prolonging the flow of current for a short period.

Inductors placed in DC circuits will initially encounter transient resistance while the magnetic field is established. Once a steady state is reached, the reactance is negligible. Conversely, inductors in AC circuits encounter increasing reactance proportional to the frequency. This is because the creation and subsequent reversal of magnetic field development produces a constant back-emf resisting current flow. Therefore, high-frequency AC will cause a high reactance in the inductor. Inductors are used to filter out high-frequency alternating currents, or to smooth out the effect of power surges in monitoring equipment.

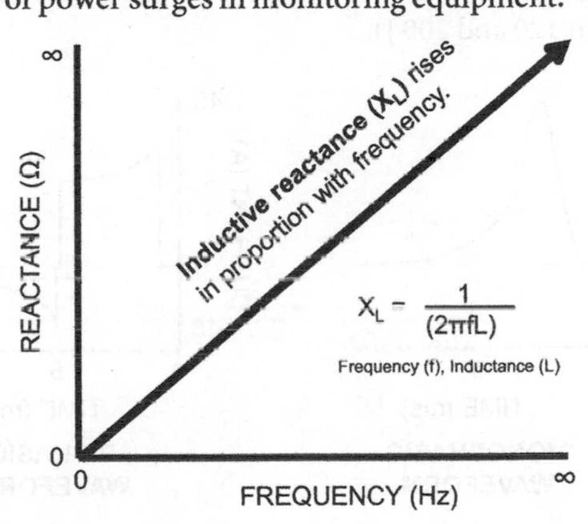

Applied science

Explain how a defibrillator works

Defibrillators deliver DC to the myocardium to treat VF and pulseless ventricular tachycardia (VT) and for synchronised cardioversion of tachy-arrhythmias. The current causes ion shifts within myocardial cells, disrupting their ion gradients, which induces a state akin to the refractory period. During this time, the heart's normal automaticity can be re-instigated, restoring sinus rhythm.

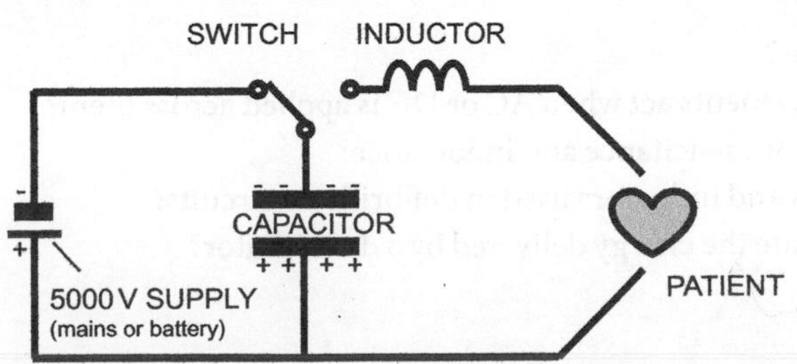

For successful defibrillation, the defibrillator must deliver energy to the myocardium over a short period. The transthoracic impedance with large surface area defibrillator pads is 50–150 Ω, so a high voltage is required to provide adequate current flow during defibrillation. A step-up transformer converts mains 240 V to 5000 V potential, and a rectifier transforms it to DC.

The 'charge' button on the defibrillator applies the 5000 V across a capacitor, which stores the charge. This enables the energy to be delivered rapidly when required. Pressing the 'shock' button causes the charge to pass through an inductor before it is finally delivered to the patient. The value of the inductor governs the duration for the steady state flow of current.

The amount of energy delivered by the defibrillator can be calculated by:

$$Energy\ (E) = \frac{1}{2} \times Capacitance\ (C) \times Voltage^2\ (V^2)$$

What is the difference between monophasic and biphasic waveforms?

Defibrillators traditionally used a monophasic sine pattern waveform in which current flows in only one direction. This delivers between 200 and 360 J to the myocardium. In the late 1990s, biphasic waveforms, which use a bi-directional current, were shown to be more successful at achieving defibrillation. This was despite the fact they used lower voltages and delivered less energy (between 120 and 200 J).

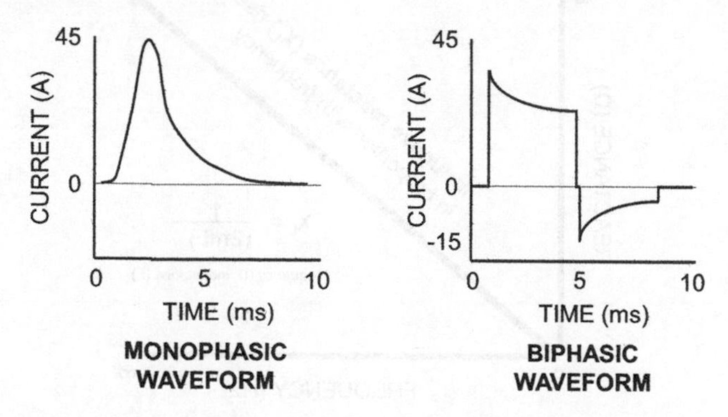

The reason for biphasic superiority remains to be elucidated. One theory suggests that the peak current flows are lower with biphasic waveforms. This prevents excessive ion translocation across cell membranes in the tissues (the mechanism by which electric shock causes damage). Another theory suggests that the initial wave polarises electrodes and skin, which reduces impedance to the subsequent wave, allowing more effective and rapid energy delivery to the myocardium.

Questions

What is a capacitor?

What is an inductor?

How do these components act when AC or DC is applied across them?

What are the units of capacitance and inductance?

Why are capacitors and inductors used in defibrillator circuits?

How do you calculate the energy delivered by a defibrillator?

Chapter 22

Diathermy

Basic science

Diathermy is an electrosurgical tool used to cut or coagulate tissues.

Cutting mode employs a continuous wave of high-frequency alternating current (0.5–1 MHz), applied via a fine electrode. The small area of the electrode results in a high **current density** at the point of contact. Energy contained within the diathermy is concentrated, causing heating with ionisation and vaporisation of water in the tissues. This leads to vapour expansion and tissue fragmentation, which cuts tissues.

In coagulation mode, the wave is intermittent (pulsed) and damped, so less energy is transmitted. The threshold for vapour expansion is not reached; therefore, tissues are denatured rather than destroyed.

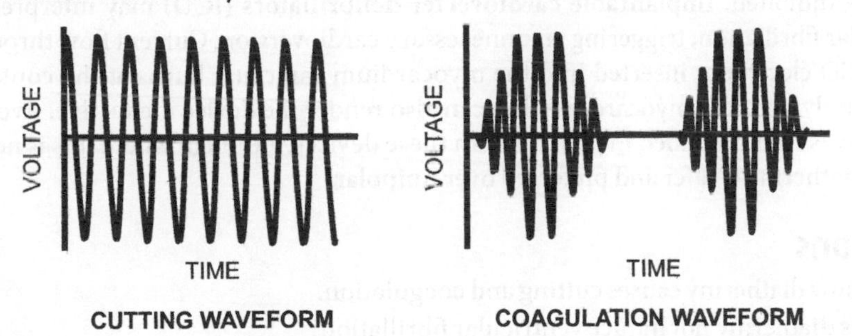

With **unipolar diathermy**, electric current flows through the patient and out via a large conductive pad placed on a nearby limb. The large surface area ensures that current density remains low, preventing burns at this site.

Chronaxie is the minimum time required for a constant electric current, which is double the threshold strength needed to stimulate a specific muscle or neuron. It is directly related to the density of voltage-gated sodium channels within the tissue, as these channels are responsible for phase 0 of the action potential and therefore the excitability of the cell. The myocardium has a chronaxie of three milliseconds. The high frequency of diathermy ensures that it does not exert a prolonged constant current, making it unlikely to trigger ventricular fibrillation.

Bipolar diathermy uses two closely positioned electrodes, such as the two tips of forceps. Current flows from one electrode through nearby tissues to the other electrode, completing the circuit. Bipolar diathermy is used in neurosurgery or eye surgery where delicate control is needed, or where current flow through the body may be harmful, for example when pacemakers are *in situ*.

Applied science

What risks are associated with diathermy use?

Diathermy uses an electrical current; all dangers associated with electrical devices are applicable. To prevent mains electricity from causing shock, diathermy uses an isolating capacitor, which permits high-frequency currents, but resists low-frequency AC.

Specific risks include burns, and monitoring and pacemaker interference.

Improperly positioned conducting pads or inadvertent patient contact with an earthed object may result in high current density at the contact site, causing **burns**. Small cross-sectional areas such as digits or metallic implants may cause channelling of the current, which can result in heating and burns. Operator error can also unintentionally damage surrounding structures, causing injury.

Arcing causes fires and explosions of flammable or combustible material. This occurs when air between the electrode and the skin becomes ionised by the strong charge at the electrode, and is able to carry current across the gap.

Monitoring, such as the electrocardiogram (ECG), detects and analyses subtle changes in skin potentials. Diathermy creates significant electrical noise, interfering with this equipment. Insulated wires, bandwidth filters and post-processing software can reduce the amount of interference.

Implantable pacemakers may interpret diathermy noise as cardiac electrical activity and be falsely inhibited. **Implantable cardioverter defibrillators** (ICD) may interpret noise as ventricular fibrillation, triggering an unnecessary cardioversion. Current flow through pacemaker/ICD electrodes inserted into the myocardium can cause burns at the contact point. This not only scars the myocardium, but can also render the device ineffective. Avoidance of diathermy is recommended in patients with these devices, although where this is not feasible bipolar diathermy is safer and preferred over unipolar.

Questions

Explain how diathermy causes cutting and coagulation.

Why does diathermy not induce ventricular fibrillation?

Describe the differences between unipolar and bipolar diathermy.

Chapter 23

Wheatstone bridge and strain gauges

Basic science

The **Wheatstone bridge** is an electrical circuit that uses an arrangement of four resistors to measure an unknown electrical resistance.

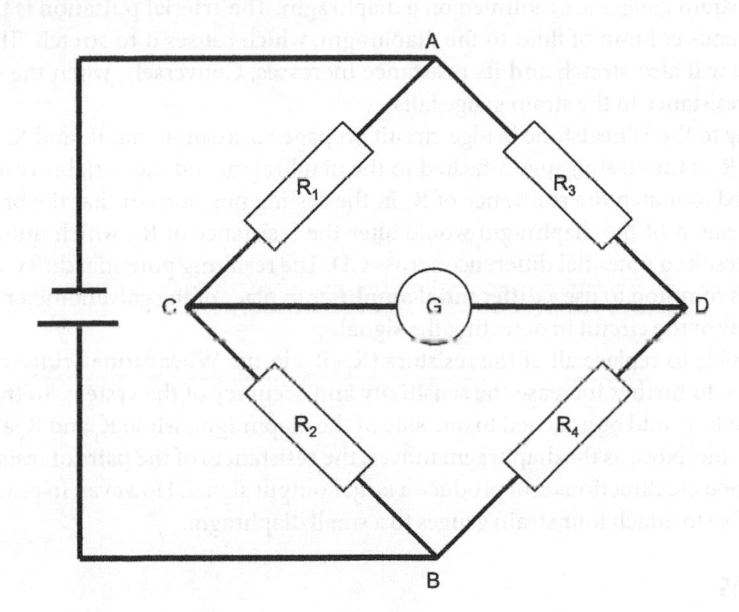

The typical Wheatstone bridge arrangement as shown above contains a power source, a galvanometer (G), two resistors of known resistance (R_1, R_2), a variable resistor (R_4) and an unknown resistance, which is the one to be measured (R_3). The connection across CD containing the galvanometer is known as the bridge.

This circuit is sensitive to changes in the **ratio of resistances** across pairs of resistors. When the voltages at C and D are equal, the ratios of resistances equal each other ($R_1/R_2 = R_3/R_4$), no current will flow through the galvanometer and the bridge is balanced.

$$\frac{R_1}{R_2} = \frac{R_3}{R_4}$$

In the circuit above, R_3 is the unknown resistance to be measured. The ratio of resistors R_1 and R_2 (ACB) is known. If the bridge is unbalanced, there will be a voltage present across CD and current will flow through the galvanometer. By altering the resistance of the variable resistor R_4 until the ratio of resistance across the limb ADB equals that of ACB, the bridge can be balanced, and no current flows across CD. By knowing the resistance required at R_4 to balance the bridge, R_3 can be calculated by using the equation above.

Applied science

What principles underlie the function of the pressure transducers found in arterial lines?

Pressure transducers are based on the principle that electrical resistance changes with pressure.

A **strain gauge** is either a foil arrangement or a conductive metallic strip. In the arterial transducer, strain gauges are mounted on a diaphragm. The arterial pulsation is transmitted via a continuous column of fluid to the diaphragm, which causes it to stretch. The attached strain gauge will also stretch and its resistance increases. Conversely, when the diaphragm relaxes the resistance in the strain gauge falls.

Referring to the Wheatstone bridge circuit on page 55, assume that R_1 and R_2 have equal resistances. R_3 is the strain gauge attached to the diaphragm, and the variable resistor R_4 has been adjusted to match the resistance of R_3 in the resting position, so that the bridge is balanced. Movement of the diaphragm would alter the resistance of R_3, which unbalances the bridge and results a potential difference across CD. The resulting potential difference is quite small, so it is common to use a **differential amplifier** in place of the galvanometer to increase the sensitivity of the circuit in detecting the signal.

It is possible to replace all of the resistors (R_1–R_4) in the Wheatstone bridge circuit with strain gauges, to further increase the sensitivity and accuracy of the system. In this arrangement, R_1 and R_2 would be attached to one side of the diaphragm, while R_3 and R_4 are attached to the other side. Now, as the diaphragm moves, the resistances of the pairs of resistors would change in opposite directions and produce a larger output signal. However, in practice it may not be possible to attach four strain gauges to a small diaphragm.

Questions

What is a Wheatstone bridge circuit?
Give examples of its clinical application.
What is a strain gauge?
How does a strain gauge function?
Why is the Wheatstone bridge often used with a differential amplifier instead of a galvanometer?

Chapter 24

Biological signals

Basic science

The movement of ions across cell membranes during the depolarisation and repolarisation of myocytes and neurones generates electric potentials. Silver metal **electrodes** covered with a layer of silver chloride gel within an adhesive sponge pad can be used to measure these potentials at the skin. Ion movement near the electrode–skin interface induces movement of chloride ions within the gel layer. The ion concentration gradient promotes electron production at the electrode.

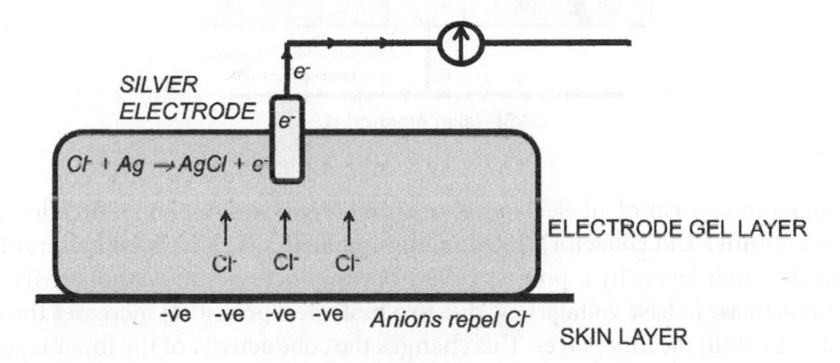

A lead wire and voltmeter attached to the electrode allows measurement of the potential relative to a reference point. The reference point is usually a second skin electrode. Signals are then amplified, processed and displayed.

Skeletal and cardiac muscles have higher amplitudes than cerebral neurones. This is because the **amplitude** of biological potentials is proportional to the number of simultaneously depolarising cells. The **frequency** of potentials is related to the fluctuating ion activity across cell membranes. Skeletal myocytes which undergo tetany have high frequencies of up to 1 kHz. Conversely, cardiac myocytes have lower frequencies due to their refractory periods. Typical amplitude and frequency ranges for the ECG, electromyogram (EMG) and electroencephalogram (EEG) are shown below.

	Amplitude (mV)	Frequency (Hz)
ECG	0.05–3	0.01–150
EMG	0.001–100	50–3000
EEG	0.001–1	δ 1–4
		θ 4–7
		α 7–14
		β 14–60

Bioelectric potentials are attenuated as they pass through tissues; a 90 mV ion potential measured at the heart is reduced to 2 mV at the skin. **Amplifiers** are electronic circuits that increase the voltage of signals prior to further processing and/or display. They contain **transistors** which are used to amplify signals and are made of semi-conductors such as, silicon, germanium or gallium arsenide.

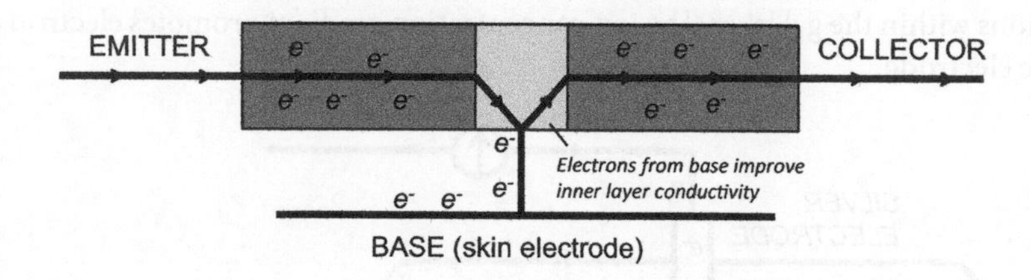

The transistor consists of an electron-poor inner layer sandwiched by two electron-rich outer layers (emitter and collector). Introducing impurities such as boron aluminium and gallium to the outer layers by a process called **doping** increases the conductivity of these regions. An increase in base voltage (e.g. due to a biological potential) increases the electron concentration within the inner layer. This changes the conductivity of the inner layer, allowing modulation of the greater current through the collector, hence amplifying the base signal.

The **gain** of an amplifier is the ratio of the output voltage or power from the amplifier to the input voltage or power, respectively. The **decibel** is a logarithmic scale that is used to describe the gain of amplifiers.

$$Amplifier\ Gain\ (decibels) = 20\ log \frac{Amplifier\ Output\ Voltage}{Amplifier\ Input\ Voltage}$$

$$Amplifier\ Gain\ (decibels) = 10\ log \frac{Amplifier\ Output\ Power}{Amplifier\ Input\ Power}$$

The **bandwidth** of an amplifier describes the range of frequencies over which it functions accurately. **Frequency response** of an amplifier is a measure of its ability to replicate an enhanced copy of the input signal without distortion of the frequency and phase.

Operational amplifiers are integrated circuits used to amplify the difference between two input signals, known as **differential amplification**. They comprise several transistors, resistors, capacitors and diodes. A simplified diagram is shown.

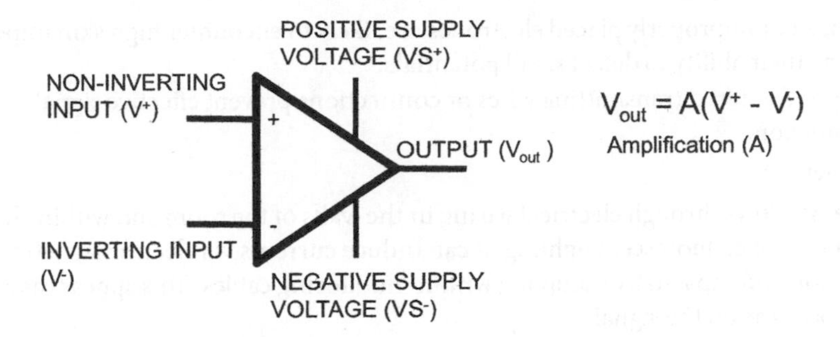

The amplifier receives input from two sources (V^+ and V^-), for example a measuring electrode (V^+) and a reference electrode (V^-). High input impedance reduces the attenuation of the input signals. It also maximises the voltage difference across the device, which is amplified. Another consequence of the high input impedance is the prevention of excessive leakage currents flowing through the circuit, minimising unwanted signal noise. A positive and negative power supply (VS^+ and VS^-, respectively), known as a **split supply**, determines the output voltage of the amplifier. Power supply voltages range from 6 to 18 V.

Signals can be displayed in either analogue or digital form. **Processing** involves converting a continuous **analogue** signal to a **digital** one by sampling the analogue signal at regular intervals.

Fourier analysis breaks down complex waveforms into simpler component sine waves. This allows analysis of the individual components of the signal, such as determining the different waveform frequencies within the EEG, or filtering out noise.

Display of analogue signals most commonly occurs via an integrated electronic moni tor, which can exhibit several signals simultaneously. Examples of display devices include oscilloscopes and printers.

Applied science

What sources cause interference when measuring biological signals?

The relatively small amplitudes of biological signals make them susceptible to distortion and error from interference created by the patient, equipment and the environment.

Patient:

- **Movement artefacts** occur due to stretching of skin beneath the electrode. This creates small electrical potentials within the epidermis, which are detected by the electrode. These appear as baseline shifts of the signal.
- **Other bioelectric potentials**, e.g. pectoral muscle contraction, create potentials that are picked up by ECG chest leads.

Equipment:

- **Electrode offset voltages** appear when the electrode stores or creates its own potential (**half-cell potential**). This interferes with its ability to detect skin potentials. Silver/silver chloride electrodes have a relatively low half-cell potential of 220 mV, making them more reliable than some older designs. Exposure to high voltages such as those used in defibrillation can temporarily polarise the electrode, creating this type of interference.

- Damaged or improperly placed electrodes are likely to encounter high skin impedance, limiting their ability to detect small potentials.
- Faulty or fractured transmitting wires or connections prevent effective signal transmission.

Environmental:

- **50 Hz AC** flows through electrical wiring in the walls of the room and within electrical equipment, e.g. fluorescent lighting. It can induce currents via electromagnetic induction and capacitative coupling within monitoring cables. This appears as a fuzzy baseline on the signal.
- Use of equipment nearby, for example **diathermy**, causes interference by direct transmission of current from the probe to the electrode, radio transmission from the wire connecting the probe to the diathermy machine, and feedback via the mains supply to the display monitor.

How can interference be minimised when measuring bioelectric potentials?

General measures to reduce interference include the following.

- Ensuring patients remain still (where possible).
- Placing electrodes on hairless, clean and dry skin to improve contact and reduce impedance.
- Annual servicing and maintenance of equipment to detect micro-fractures in wires and connections.
- Avoiding devices that produce signals with frequencies within the range of those being measured.

Some features of the equipment design can also reduce interference.

- **Insulated wires** are surrounded by a dielectric material within a casing of conductive material, known as a **Faraday's shield**. In the presence of external interference, e.g. a magnetic field or capacitative coupling, impulses are formed in the surrounding insulating material and conducted away, protecting the enclosed wires.
- **Twisted pair cabling** contains two entwined wires holding equal and opposite signals from the same electrode. At the destination, one of the signals is inverted and the two inputs are combined using an operational amplifier. Because interference will affect both wires equally along their route, signal that is common to both wires will be cancelled out. This process is known as **common mode rejection**.

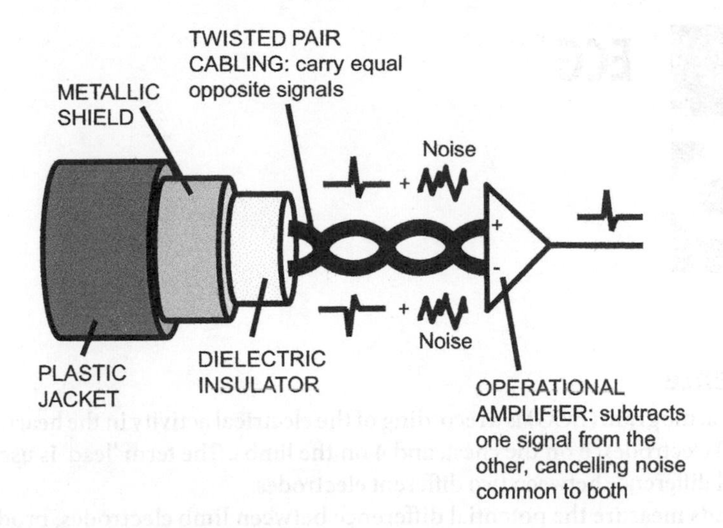

How can processing improve signal quality?

Processing serves to increase the **signal to noise ratio** (the ratio of signal amplitude to noise amplitude, expressed in decibels). This can be achieved using differential amplifiers, filtering and averaging.

Differential amplifiers reduce interference by comparing potentials from a measuring and a reference electrode by using common mode rejection. For the ECG, an electrode placed on the patient's right leg often serves as the reference electrode. This electrode is connected to a specialised circuit known as a '**right leg drive**' that acts to actively reduce noise interfering with the measuring signal.

Filtering unwanted frequencies removes noise from the desired signal. **Low-pass filters** filter out high-frequency signals (such as diathermy signals), while **high-pass filters** will filter out low-frequency signals. **Band pass filters** allow frequencies within a determined frequency band to pass. For example, a 0.5–100 Hz band filter would be suitable for monitoring ECG signals. **Notch filters** can be used to reject a specific frequency, such as the 50 Hz frequency of mains electricity.

Averaging involves taking multiple measurements of the same signal, and using their mean value. Noise is assumed to be random, and has a mean of 0 and a constant variance; it therefore cancels itself out over repeated measurements.

Questions

Where do biological potentials originate?
What types of biological signals do we measure?
Compare and contrast their frequencies and amplitudes.
Describe the process of analysing and displaying biological signals.
What are electrodes made of and how do they function?
Why are amplifiers necessary?
Explain common mode rejection.

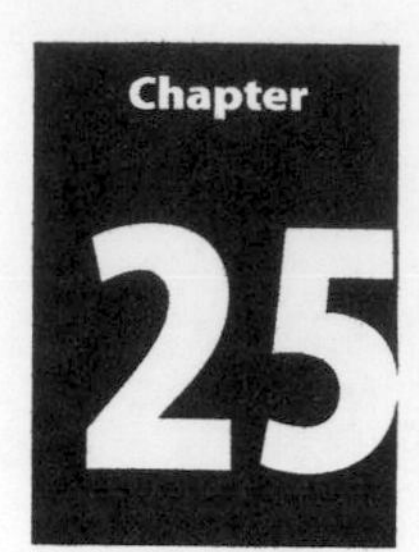

ECG

Basic science

The **electrocardiogram** (ECG) is a recording of the electrical activity in the heart. The 12-lead ECG uses 10 electrodes, 6 on the chest, and 4 on the limbs. The term 'lead' is used to refer to the potential difference between two different electrodes.

Limb leads measure the potential difference between limb electrodes, producing leads I, II and III as shown by Einthoven's triangle. **Augmented leads** (aVR, aVL and aVF) show the potential difference between each limb electrode and **Wilson's central terminal.** The potential at Wilson's central terminal is derived from the average potential of the three limb electrodes. **Chest leads** (V1 to V6) also use Wilson's central terminal as their reference point. The right leg electrode forms part of the **right leg drive** circuit which is used for noise reduction.

The placement of ECG electrodes for the 12-lead ECG, and Einthoven's triangle are shown below.

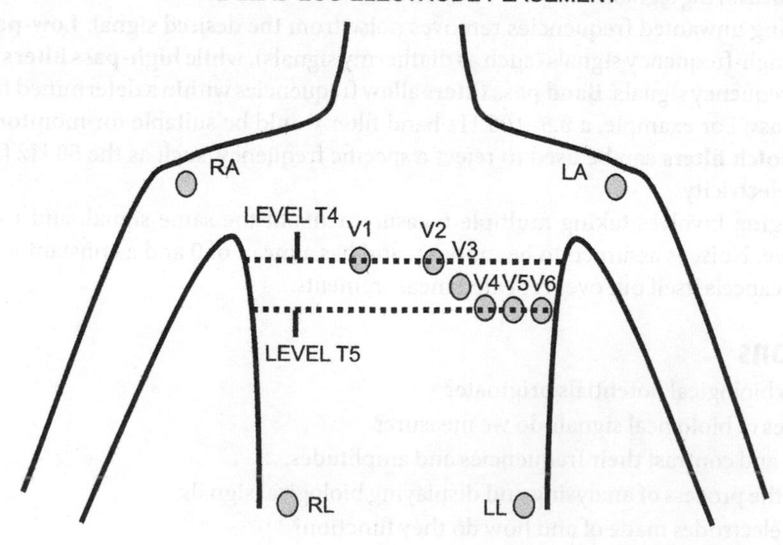

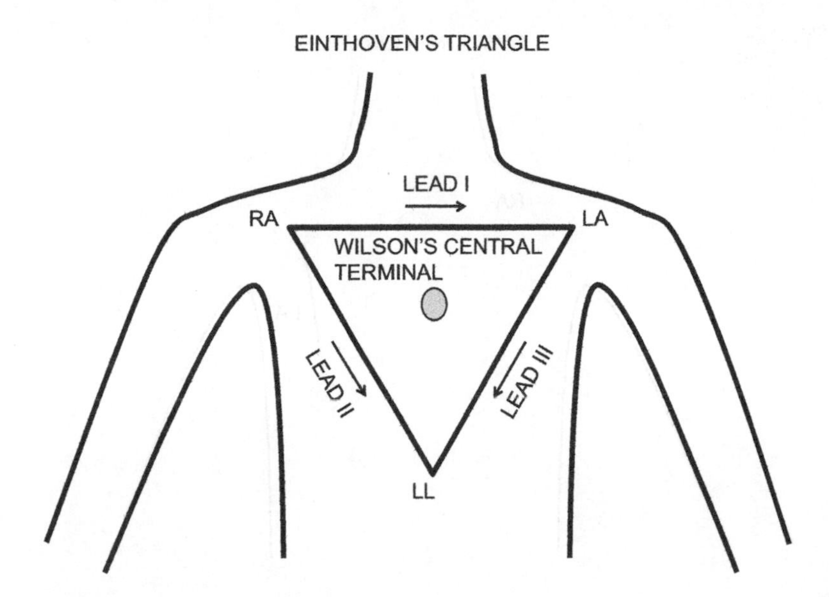

Cardiac impulses begin in the right atrium at the sino-atrial node. The wave of depolarisation spreads through both atria to the atrioventricular node. Here it travels via the bundle of His and Purkinje fibres to innervate the ventricular muscle. The general direction of depolarisation forms the cardiac axis.

Deflections of the ECG depend on the position of the electrode in relation to the direction of depolarisation. Positive deflections indicate depolarisation moving towards the electrode; leads that lie parallel to the cardiac axis have predominantly positive QRS complexes (e.g. lead II). Negative deflections indicate depolarisation moving away from the electrode.

Applied science

How is the ECG used in anaesthetic practice?

The ECG is regarded as an essential monitor during anaesthesia. It provides vital information such as heart rate and rhythm, and can indicate myocardial ischaemia.

High- and low-pass filters can be selected for use with the ECG depending on the intended use. In **diagnostic mode**, where a greater diagnostic accuracy for ST segments is required, the high-pass filter is set at 0.05 Hz and the low-pass filter at 40, 100 or 150 Hz. For routine rhythm monitoring during anaesthesia, where noise can be problematic, a more filtered **monitoring mode** is used. In monitoring mode, the high-pass filter is set higher to 0.5 Hz and the low-pass filter at 40 Hz.

A full 12-lead ECG is unwarranted during routine anaesthesia, as the limb leads alone provide enough information. An alternative configuration of the limb leads is the **CM5 Configuration** (Central Manubrium V5). In the CM5, the right arm electrode is placed at the suprasternal notch, the left arm electrode at V5 position, and the left leg electrode on the left shoulder or leg to act as a reference point. This configuration lies parallel to the axis of the left ventricle, and is useful for detecting left ventricular ischaemia.

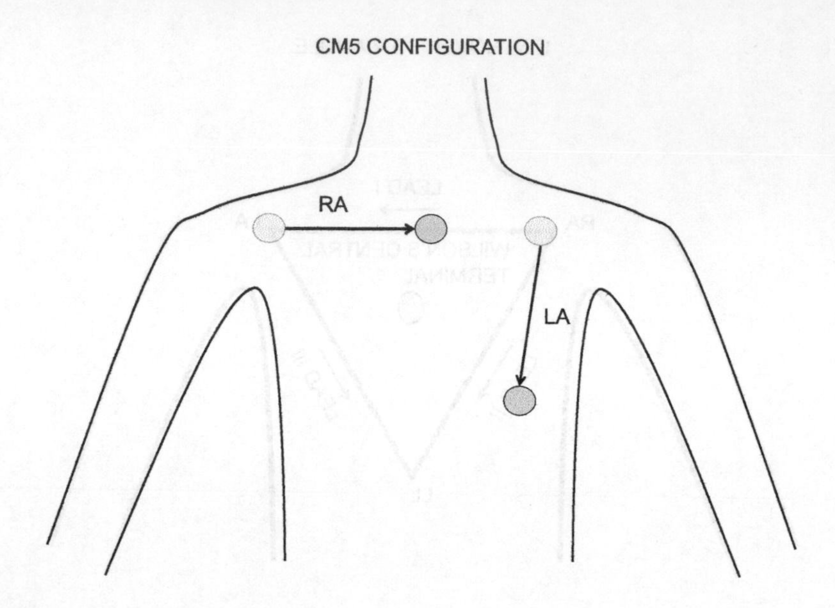

Questions

What is meant by the term lead in ECG?

Describe the lead position of the 12-lead ECG.

Draw Einthoven's triangle.

What is the difference between diagnostic and monitoring ECG modes?

Describe the CM5 position of ECG leads. Why it is used?

Chapter 26

EEG and EMG

Basic science

The **electroencephalogram (EEG)** records brain activity by using 20–22 scalp electrodes. Potentials produced by cerebral neurones are displayed as waves. EEG frequency components include:

Wave	Frequency (Hz)	Interpretation
δ	1–4	Normal in children and during sleep
		Indicates disease in adults
θ	4–7	Normal in children
		Indicates disease in adults
α	7–14	Present in the occipito parietal region, increases with eyes closed and falls with complex cognitive tasks
β	>14	Present in the frontal area when awake

General anaesthesia produces a progressive increase in low-frequency, high-amplitude activity on the EEG as the level of general anaesthesia deepens. **Burst suppression** describes a characteristic repetitive pattern of slow-wave, high-amplitude activity followed by inactivity (flat EEG). This occurs in comatose states, but can also occur during deep anaesthesia.

The **electromyogram (EMG)** records the electric potentials produced by skeletal muscle contraction, either from skin electrodes or intramuscular needle electrodes. The needles are commonly made of stainless steel or platinum, and are insulated along their length, except for the tip. A skin electrode placed on an electrically neutral body part serves as the reference point. An alternative needle electrode is the **concentric needle electrode** (shown below), which has a hollow shaft, containing an insulated wire. The shaft acts as a reference electrode, and the wire tip as the measuring electrode. These are less susceptible to noise when used with a differential amplifier, although the shaft may be affected by superficial muscle contractions.

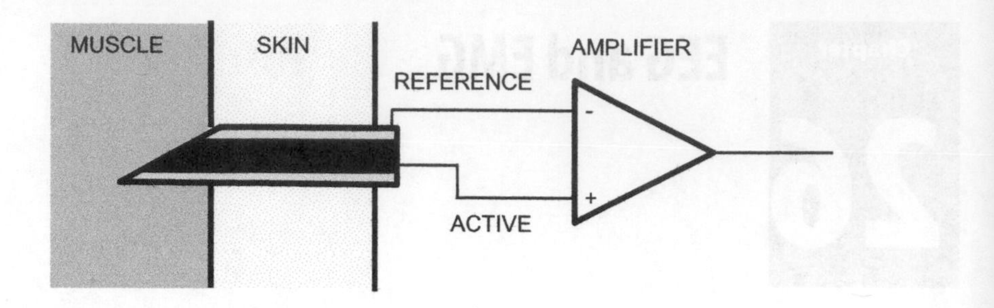

Applied science

How are the EEG and EMG used in anaesthesia?

Performing a complete EEG with 20–22 electrodes in the intraoperative environment is not practical and the raw EEG is complex to interpret. Modified EEG-based depth of anaesthesia monitors such as bispectral (BIS) analysis and compressed spectral array (computer-processed EEG) are more commonly used in the intraoperative setting. The BIS monitor analyses EEG data obtained from four electrodes placed on the forehead. This involves Fourier's analysis to determine the frequency components of the waveform, with subsequent comparison of the phase of each wave relative to others. A dimensionless number between 0 and 100 is displayed, 100 indicating fully awake and 0 the absence of cortical activity. A suggested suitable depth for general anaesthesia lies within the 40–60 range.

The EMG can be used to distinguish between disorders of muscle, nerve, and the neuromuscular junction.

Monitoring **motor-evoked potentials** during spinal surgery can be used to alert the surgical team of impending spinal cord compromise. Scalp electrodes stimulate the motor cortex at regular intervals, and the EMG response is measured through needle electrodes in the tibialis anterior, abductor hallucis or vastus medialis. Reduction in EMG activity indicates potential compromise of the corticospinal tract.

EMG can be used to assess the presence of residual block following neuromuscular blockade. A supramaximal stimulus with a unipolar square waveform current of 20–60 mA for 0.2–0.3 ms is applied to a peripheral nerve (commonly the facial nerve or ulnar nerve). The EMG response at the corresponding muscle is then measured. The EMG is considered the gold standard when assessing neuromuscular blockade, as it is able to detect muscle action potentials even when mechanical contraction is undetectable.

Questions

Which EEG frequency components are pathological?

Describe a depth of awareness monitor that you have used.

Explain the principles behind intraoperative spinal cord monitoring.

Chapter 27 Resonance and damping

Basic science

The **natural frequency of a system** refers to the number of oscillations per unit time, if that system is set into motion by an external force in a vacuum. This is the frequency at which the system will naturally oscillate if there is no external force or resistance.

The **fundamental frequency** is the lowest frequency at which a system will oscillate freely. **Harmonics** are integer multiples of the fundamental frequency.

If periodic driving impulses cause a system to oscillate, the forced oscillations occur at the same frequency as the periodic impulses and have constant amplitude. This is because the energy causing the oscillations is supplied by the driving impulses. **Resonance** occurs when the frequency of the periodic driving impulses matches the natural frequency of the system. The addition of energy from the periodic driving impulses is timed such that they add to the existing oscillations, causing increase in amplitude with each subsequent oscillation. The oscillations will continue to grow in amplitude until maximal amplitude is reached or the system is destroyed.

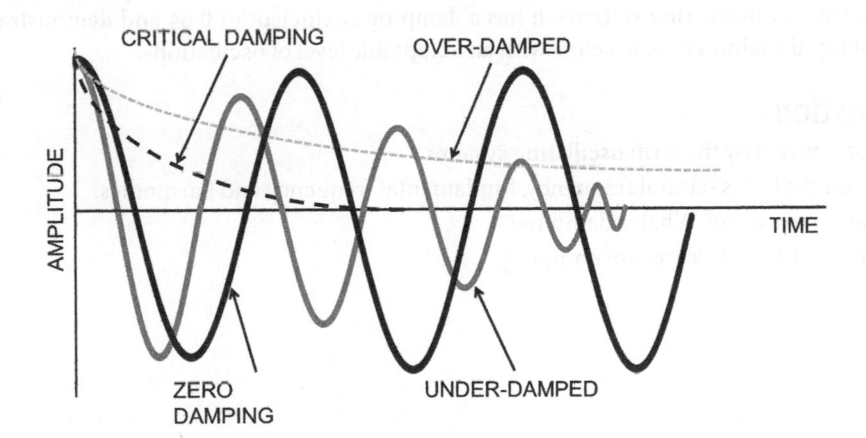

Damping results in oscillations of reducing amplitude due to the loss of energy to the surroundings. It is measured as the **damping coefficient,** which refers to the amount of damping present in a system. A damping coefficient of zero indicates no loss of amplitude with time, critical damping occurs at a damping coefficient of one, and an over-damped system has a damping coefficient of greater than one.

Free oscillations occur when there is no loss of energy to the surroundings (**zero damping**) and this is seen by oscillations of the same amplitude. On the other hand, a **critically damped** system will fail to show any oscillations following an impulse but it has the fastest return to baseline without oscillating. An **over-damped system** (damping coefficient > 1) will fail to oscillate and have a very slow return to baseline following an impulse.

Applied science

Why are resonance and damping important when measuring invasive blood pressure?

Invasive blood pressure measurement uses a cannula connected to a saline column and pressure transducers to convert an arterial pressure waveform into an electrical signal.

The resonant frequency of the measuring system needs to be outside the range of frequencies present in the blood pressure waveform to avoid resonance, which distorts the displayed pressure waveform. The fundamental frequency is the heart rate and the first 10 harmonics contribute to the pressure waveform. Therefore, the resonant frequency of the measuring system needs to be at least 10 times the fundamental frequency. Using short non-compliant tubing, the fundamental frequency of the system can be raised, and this reduces resonance.

The loss of energy to the surroundings, as a result of damping, will affect the speed at which an invasive pressure measuring system is able to respond to the changes in arterial pressure.

An over-damped system results in a blood pressure waveform that is excessively blunt and slow to return to baseline. An under-damped system shows excessive oscillations following each pressure wave. Both make interpretation of systolic and diastolic pressure inaccurate.

A critically damped arterial pressure waveform shows no oscillations before returning to baseline, but the response time will be relatively slow. For this reason, **optimal damping** is preferred for measuring systems. It has a damping coefficient of 0.64 and demonstrates the most rapid attainment of baseline with an acceptable level of oscillations.

Questions

What is meant by the term oscillating system?

Explain the terms natural frequency, fundamental frequency and harmonics.

What is resonance? What is damping?

Describe different degrees of damping.

Chapter 28 Ultrasound and Doppler

Basic science

Ultrasound describes sound waves oscillating at a frequency greater than 20 kHz, which is well above the range audible to the human ear. Clinical ultrasound uses frequencies between 1 and 20 MHz.

Sound waves are propagated through a medium by the vibration of molecules. Sound waves cause regular pressure variations with alternating areas of **compression** (high pressure) and **rarefaction** (low pressure), which are depicted in the following diagram.

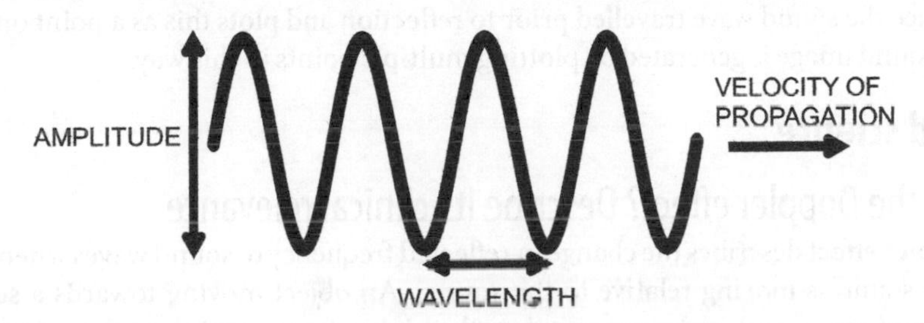

Wavelength is the distance between two points of maximum compression or rarefaction. The **frequency** of a wave is the number of wavelengths per second expressed as **hertz (Hz)**. High-frequency sounds have shorter wavelengths than low-frequency sounds. Interaction with an object is dependent on the wavelength being the same size or smaller than the object. Higher-frequency waves provide **better resolution** (higher quality) by interacting with smaller objects, but have lower penetration due to rapid attenuation. Conversely, lower frequencies penetrate further at the expense of resolution.

The **amplitude** of a wave is the signal intensity, or loudness of the sound, expressed in decibels.

Propagation velocity is the speed at which sound moves through a medium. It depends on the density and compressibility of the medium. Different tissues such as blood, bone, fat, lung and muscle have different propagation velocities. Taking into account these differences, the average speed at which ultrasound passes through human tissues is 1540 $m \cdot s^{-1}$ at 37°C.

Ultrasound waves passing through tissues can undergo reflection, refraction, attenuation and scatter.

Reflection: ultrasound waves are partly reflected at the boundaries of tissue interfaces. These reflected sound waves form the basis of ultrasound imaging. The amount reflected depends on the difference in acoustic impedance between two interacting tissues. The **law of**

reflection states that the angle of incidence equals the angle of reflection. Therefore, interfaces that are perpendicular to the beam will reflect waves back to the probe to the greatest extent.

Refraction: bending of the ultrasound beam occurs when it encounters tissues of differing acoustic impedance at oblique angles.

Attenuation: as the wave passes through tissues, it causes molecules to vibrate, leading to energy loss as heat. This results in diminution of the signal.

Scatter: small structures such as blood cells cause some of the ultrasound beam to scatter, contributing to signal attenuation.

Ultrasound transducers use **piezoelectric crystals** to generate and receive ultrasound. Applying an electric current to a piezoelectric crystal aligns the polarised particles with the crystal surface, resulting in a change in shape of the crystal. An alternating current causes rapid expansion and contraction of the crystal, which produces compressions and rarefactions (i.e. a sound wave). Using the same principle, the piezoelectric crystal can receive incoming sound waves to create an alternating electrical current.

Ultrasound images are created by the detection and subsequent display of reflected ultrasound waves. Piezoelectric crystals in the transducer generate a brief pulse of ultrasound waves, and then enter a receiving mode where they detect reflected ultrasound waves. Assuming the average speed of sound in human tissues is 1540 $m \cdot s^{-1}$, the machine calculates the distance the sound wave travelled prior to reflection and plots this as a point on a screen. An ultrasound image is generated by plotting multiple points in this way.

Applied science

What is the Doppler effect? Describe its clinical relevance

The **Doppler effect** describes the change in reflected frequency of sound waves when an object reflecting sound is moving relative to the source. An object moving towards a source will reflect sound waves with a shorter wavelength (higher frequency) than the original signal. This occurs because each incoming wave is intercepted and reflected sooner by an object moving towards the source. Conversely, objects moving away from a source reflect waves with a lower frequency. This change in frequency is known as the **Doppler shift**.

Doppler allows assessment of the direction and velocity of blood flow. When ultrasound waves pass through a blood vessel, they interact with red blood cells and some waves are reflected back to the transducer. The Doppler shift produced by the movement of red blood cells is calculated using the equation below.

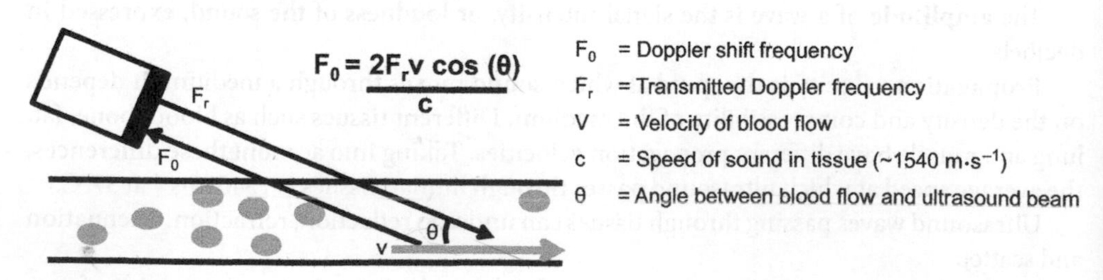

This equation highlights the importance of correct probe alignment when calculating the Doppler shift. The most accurate calculations occur with θ angles as close to 0° as possible (cosine 0° = 1). Therefore, the angle between the direction of blood flow and the ultrasound signal should be kept to a minimum.

Doppler cardiac output monitors measure red blood cell velocity in the aorta. A velocity–time graph can be generated from the Doppler flow signal. The area under this graph represents the **velocity time integral** (VTI), which represents the distance travelled by red blood cells (**stroke distance**). The product of VTI and the aortic cross-sectional area gives the volume of blood ejected per beat (**stroke volume**). The cardiac output can then be deduced by multiplying the heart rate by the stroke volume.

Describe some causes of inadequate ultrasound images

Excessive **gain** results in images that appear excessively white with poorly defined structures. Conversely, gain set too low results in a dark image and some structures may not be displayed.

Diathermy uses high-frequency current and causes **electrical interference**, which creates artefact. This appears as a series of rippled lines across the image.

Acoustic shadowing occurs where tissues distal to a highly reflective structure are poorly visualised or absent. Tissues with high reflectivity will reflect the majority of the ultrasound signal back towards the transducer, leaving little signal to penetrate deeper structures. This causes **dropout**, which appears as a poor or absent image behind a bright structure. This is classically seen when bone is encountered.

Air between the probe and tissues being examined can significantly attenuate signals. This is because air is a relatively poor conductor of sound. Water-based gels are used to improve contact between the probe and the skin to overcome this problem. However, air beneath the skin surface may be more problematic, for example when trying to image thoracic structures lying deep to the lungs, or in the case of surgical emphysema.

Questions

What is ultrasound?
Explain the terms wavelength, frequency, amplitude and propagation velocity.
What frequency range is used for clinical imaging?
What is the average speed of ultrasound in human tissues?
What is a piezoelectric crystal? How are they used in transducers?
Describe how ultrasound waves interact with tissues.
What is Doppler shift?
What is a velocity time integral?
How is this used to calculate cardiac output?

Chapter 29

Light and optical fibres

Basic science

Light is the part of the **electromagnetic spectrum** that is visible to the human eye. It has wavelengths between 380 and 740 nm.

The **electromagnetic spectrum** is comprised of propagating waves produced by interactions between electric and magnetic fields. Electromagnetic waves are **transverse**, meaning that the vibrations forming the waves are perpendicular to the direction of travel. Similar to all waves, electromagnetic waves have a frequency, wavelength and speed. **Frequency** is the number of waves passing any given point per unit time. **Wavelength** is the distance over which the waveform repeats. The speed of light is therefore the product of multiplying the frequency of oscillation and wavelength.

Waves of the same frequency that reach maximal and minimal amplitude simultaneously are said to be **in phase**. In other words, their sinusoidal waveform graphs overlap perfectly.

Huygens' theory of light describes the propagation of electromagnetic radiation as spherical waveforms, called wavefronts. Each point on a wavefront acts as a secondary emitter, producing its own spherical waveform. Because each point directly ahead of the old point is in phase, the waveform is recreated. Therefore, waves of electromagnetic radiation spread out in all directions in straight lines from the point source of emission.

A simple analogy is to think of a ripple produced in water when a stone is dropped. The circular ripple created by the stone represents a wavefront. If new stones were dropped at every point on the circular ripple, new ripples would form, and the waves would spread out radially in straight lines from the point of the original stone drop.

The speed of light in a vacuum is faster than the speed of light in glass, because light interacts with the electrons in the atoms of glass. When light crosses the boundary between two mediums, there is a change in speed, which alters the direction of travel. This is known as **refraction**. Larger drops in the speed of light as it crosses boundaries will result in a greater degree of refraction. The **refractive index** is the factor by which the speed of light is reduced when passing from a vacuum to any given medium.

Applied science

What principles underlie the function of optical fibres?

Optical fibres are produced from very high purity glass such as silica glass, and each fibre is coated with cladding. The glass prevents the light signal from scattering and the cladding prevents light leaking from the sides of the fibres.

The speed of light varies depending on the medium through which it is travelling. At a boundary or interface between two mediums, the speed and direction of travel will change. The change in direction is dependent on the angle at which the light ray, known as the incident ray, meets the boundary. This angle is known as the **angle of incidence**.

The **critical angle** refers to the angle of incidence which results in a 90° refraction of light. Incident angles below the critical angle result in both reflection and refraction, whereas incident angles above the critical angle result in complete light reflection.

Total internal reflection is where all light is reflected and none is refracted. To achieve this, light must enter the fibre at an angle which is greater than the critical angle. Total internal reflection occurs at the interface between the core of the fibre and the cladding, because the cladding has a lower refractive index. This principle is used to guide light along the length of the fibre, regardless of whether the fibre is straight or bent.

Fibre-optic scopes used in clinical practice require two bundles of fibres to avoid interference. One transmits light towards the patient and the other carries the signal back up to the camera.

Questions

What is the electromagnetic spectrum?
What is light?
Define wavelength, frequency and phase.
Describe Huygens' theory of light.
Where do we use fibre-optic devices in clinical practice?
What are the physical principles behind their use?
Describe the terms angle of reflection, critical angle and total internal reflection.

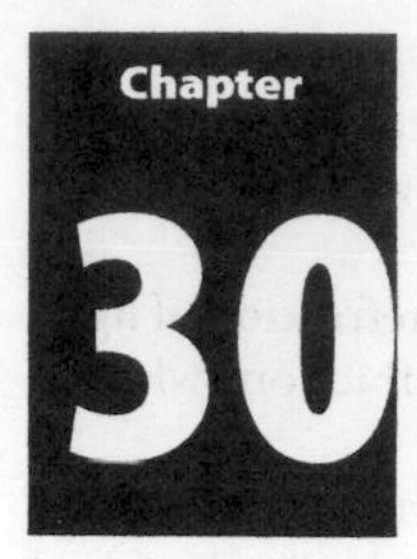

Laser

Basic science

LASER is an acronym for Light Amplification by Stimulated Emission of Radiation.

Lasers produce an intense parallel beam of **coherent monochromatic light** (one specific wavelength of light) by the stimulated emission of photons from excited atoms.

Electrons of atoms within a lasing medium normally reside in a stable low-energy level known as the **ground state**. Application of energy to atoms within the lasing medium is known as pumping. This excites electrons, raising them to higher energy levels. Because higher energy states are unstable in comparison to the ground state, there is a tendency for electrons to release excess energy and return to lower energy levels. This process is known as **decay**.

Lasing mediums can comprise multiple energy levels, such as three- or four-level systems. A three-level system is shown below.

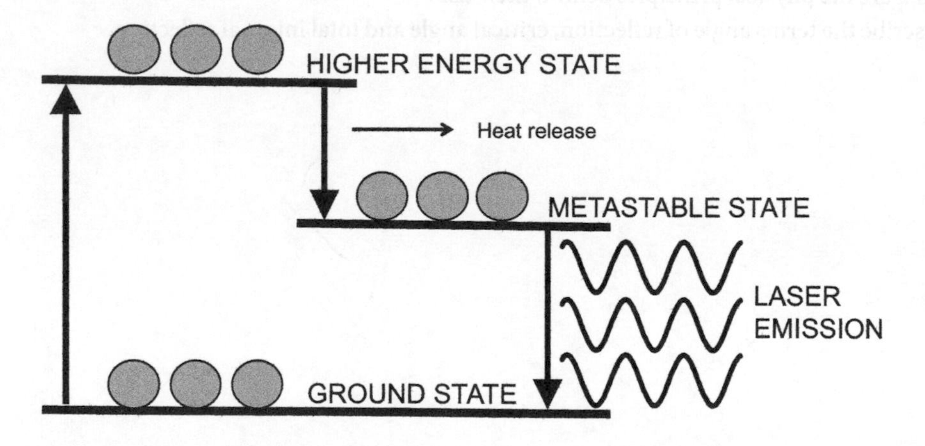

The highest energy level is the most unstable state, and electrons quickly decay by releasing energy in the form of heat to a slightly lower energy level called the **metastable state**. The metastable state is more stable than the higher energy levels, so electrons remain in this state for a longer period of time. Continued pumping of the lasing medium results in **population inversion**, where more electrons exist in the metastable state compared to those in the ground state. This non-equilibrium condition created by population inversion is essential for laser creation.

As an electron decays from the metastable to ground state, it emits a photon of energy. If this photon strikes an excited electron in the metastable state, it incites it to emit another photon, which will have the same wavelength, waveform and direction as the incident photon. They are said to be **in phase** and **coherent.**

The lasing chamber contains a lasing medium and mirrors, which reflect photons back and forth between a totally reflecting and partially reflecting mirror. The beam of light is amplified in this manner and exits the chamber through the partially reflecting mirror as an intense parallel beam of coherent monochromatic light.

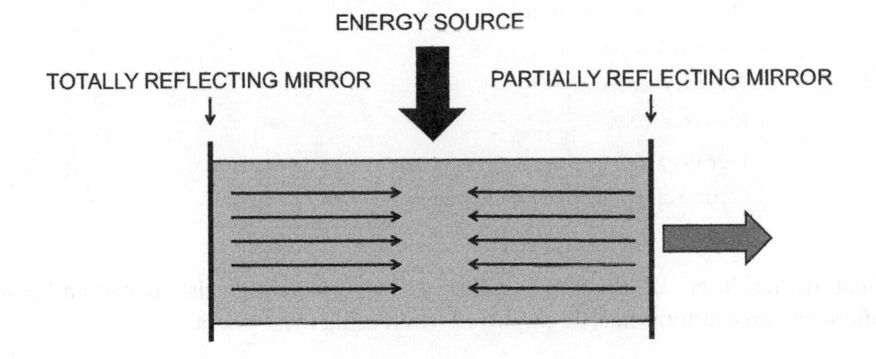

Lasers cause heating of tissues which ultimately leads to destruction. Far-infrared and ultraviolet lasers are readily absorbed by water and therefore do not penetrate deep into tissues. Lasers in the visible spectrum are absorbed by haemoglobin and melatonin and penetrate further. Red and near-infrared lasers have the deepest penetration.

Carbon dioxide lasers emit infrared light at 10 600 nm wavelength. They have limited penetration, but these lasers are precise and can be used for cutting and vaporising.

Argon lasers predominantly produce blue–green light at 488 and 514 nm, but they also emit multiple other wavelengths. They are able to coagulate tissues, and are commonly used in ophthalmology.

Neodymium-doped yttrium aluminium garnet (Nd:YAG) emits at 1064 nm. These lasers have the deepest penetration and can cut and coagulate. They are used to resect gastrointestinal and bronchial tumours, and gynaecological lesions.

Applied science

Do you know of any classification system for lasers?

In the UK, the IEC (International Electrotechnical Commission) 60825 is adopted as the British standard for laser safety classification. This classifies lasers according to their maximum output power and wavelength. The table below summarises the classes of lasers.

Class	Description
Class 1	Safe under normal usage conditions
Class 1M	Safe under normal conditions except when passed through magnifying optics
Class 2	Visible-light beam only (400–700 nm)
	Limited to power up to 1 mW
	Safe under normal conditions due to protection by blink reflex
Class 2M	As Class 2, but are **not** safe when viewing with optical instruments
Class 3R	Considered safe with careful handling and restricted viewing
	Limited to power up to 5 mW
Class 3B	Hazardous with direct exposure to eye
	Protective eye wear required
	Power up to 0.5 W
Class 4	Extremely hazardous
	Power over 0.5 W
	Hazardous with direct or scattered exposure to eyes or skin
	Capable of igniting flammable material

Most medical lasers are **class 3B and class 4.** They pose a high risk to staff and patients. Specific safety precautions must be employed when using these lasers.

What key safety issues must be considered when using lasers in theatre?

Laser safety is essential to minimise the risk posed to patients and staff. Hazards include eye injuries, burns, fires, explosions and smoke inhalation. Safety precautions include the following.

General precautions:

- Appropriate information and training for staff.
- Designation of a laser safety officer to supervise the correct use of equipment, and to ensure safety measures are followed.
- Warning signs to indicate lasers in use.
- Designated areas with restricted access and covered windows.
- Water to extinguish small non-equipment fires.

Equipment:

- Wavelength-specific goggles for eye protection.
- Protective clothing where appropriate.
- Evacuation system to extract fumes.
- Equipment used in close proximity to lasers should be fire-retardant.

Lasers can ignite flammable material such as endotracheal tubes and surgical drapes. They can also cause airway and body cavity fires in the presence of high concentrations of flammable gases.

The risk of airway fires can be reduced by using the lowest inspired oxygen concentration possible that achieves suitable oxygen saturations. In addition, using laser-safe endotracheal tubes with the cuffs filled with saline and dye helps to further reduce the risk of fires. The water in the cuff acts as a heat sink to reduce the likelihood of perforating and igniting the cuff with the laser. The dye provides a visual indication in the event of cuff perforation.

Questions

What does laser stand for?
Explain how are lasers generated.
What is meant by coherent and monochromatic light?
How does the wavelength of the laser relate to its effect on tissues?

Chapter 31 Pulse oximetry

Basic science

Light passing through a substance is partly absorbed by molecules within the substance. The more molecules present, the greater the absorbance potential. The following two laws describe the absorbance of light.

Beer's law: For a solution, absorbance is proportional to the molar concentration of the substance through which light passes.

Lambert's law: Absorbance is proportional to the thickness of the substance through which light passes, and is also related to the composition of that substance. Combining these laws:

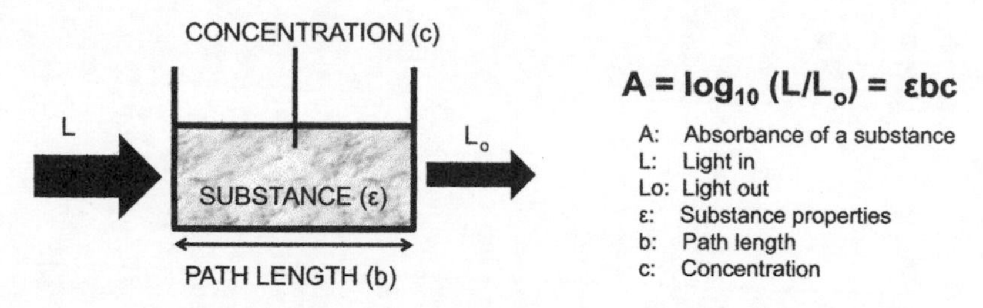

Absorbance is a dimensionless number; 0 indicates no absorbance of light, while 1 indicates 90% of light has been absorbed. Absorbance >1 is uncommon, as Beer–Lambert's law ceases to be true at extreme substance concentrations. This is due to:

- Electrostatic interactions between molecules in close proximity altering substance properties (ε).
- Scattering of light without true absorption.
- Shifts in chemical equilibria as a function of concentration.
- Changes in refractive indices at high analyte concentrations.

The pulse oximeter is a non-invasive monitor of arterial oxygen saturation, which is based upon the Beer–Lambert's law.

Transmission oximetry uses a probe placed on a finger, earlobe or foot. It contains two light-emitting diodes (LEDs), one transmitting monochromatic light at 660 nm (red) and the other at 940 nm (infrared). Light penetrates tissues and is detected by a photodiode on the opposite side of the probe, then analysed.

Oxyhaemoglobin absorbs infrared light and reflects red light (giving its red appearance), more than deoxyhaemoglobin, which absorbs more red light than infrared light. Each LED is switched on and off in turn several times a second. This allows the photodiode to analyse the absorption of each wavelength separately, because it detects the total amount of light transmitted but cannot distinguish between wavelengths. A period with both LEDs off between cycles allows adjustment for ambient light.

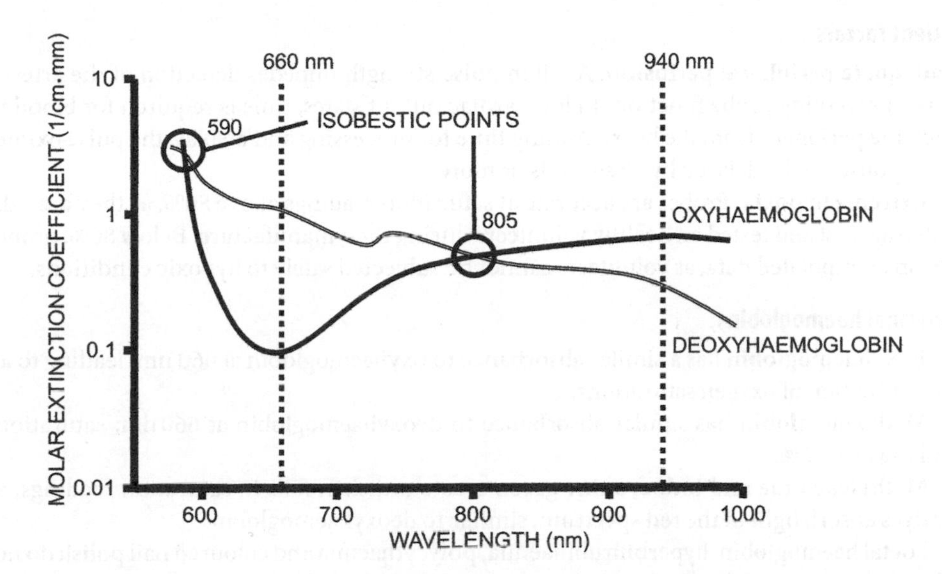

The absorbance spectra of oxyhaemoglobin and deoxyhaemoglobin are shown in the graph above. **Isobestic points** occur where absorbance is the same for both haemoglobins; this is used to calculate haemoglobin concentration.

The ratio of red to infrared light transmitted at each wavelength determines the ratio of oxyhaemoglobin to deoxyhaemoglobin, giving a saturation percentage. Venous blood, capillaries and tissues absorb a large proportion of light (depicted in the diagram below) which could provide erroneous results if analysed. Subtracting the minimum transmitted light from peak transmitted light at each wavelength allows analysis of only the pulsatile component (arterial blood) of the signal.

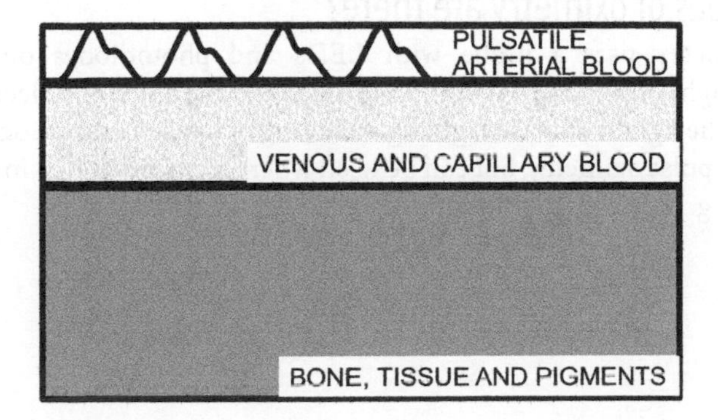

Applied science

What factors influence the pulse oximeter?

Patient and environmental factors can influence the pulse oximeter.

Patient factors

Inadequate peripheral perfusion: A fall in pulse strength impedes detection of the arterial phase, preventing probe function. In low cardiac output states, time is required for blood to reach the periphery from the heart. Adding time for processing and display, the pulse oximeter response can be delayed by 40 seconds or more.

Extreme hypoxia: Probes are accurate at saturation readings above 80%, as they are calibrated against and tested on healthy volunteers during their manufacture. Below 80%, probes rely on extrapolated data, as volunteers cannot be subjected safely to hypoxic conditions.

Abnormal haemoglobins

Carboxyhaemoglobin has a similar absorbance to oxyhaemoglobin at 660 nm, leading to an overestimation of oxygen saturations.

Methaemoglobin has similar absorbance to deoxyhaemoglobin at 660 nm; saturations tend towards 85%.

Methylene blue and indocyanine green create transient falls in saturation readings, as the dyes absorb light in the red spectrum, similar to deoxyhaemoglobin.

Foetal haemoglobin, hyperbilirubinaemia, polycythaemia and coloured nail polish do not affect the accuracy of the probe.

Environmental factors

Excessive ambient light floods the photodiode, impairing its ability to detect relatively small amounts of transmitted light. Shielded probes reduce this problem.

Movement artefact causes finger rotation, increasing the cross-sectional area and thus absorbance of transmitted light. Movement also increases ambient light entering the probe, and causes venous pulsations, which are falsely interpreted by the probe as arterial.

Electrical noise/interference, e.g. diathermy, can also affect the quality of the signal, and hence the probe's accuracy.

What other types of oximetry are there?

Reflectance oximetry uses a probe with LEDs and photodiodes on the same side. Monochromatic light of 660 and 940 nm is emitted, and the amount reflected is detected by the photodiode. The light is analysed, giving a reading of oxygen saturations. This method is less accurate than pulse oximetry, but can be used where transmission oximetry is limited by tissue thickness, e.g. chest, wrist or forehead.

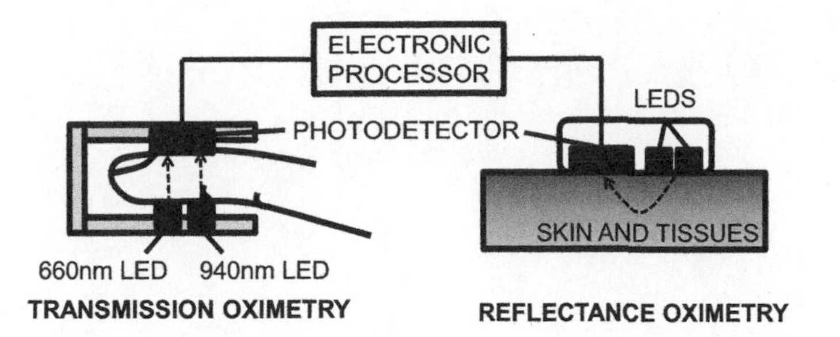

Co-oximeters found in blood gas machines provide concentrations of haemoglobin, oxyhaemoglobin, carboxyhaemoglobin and methaemoglobin. Several wavelengths of light are passed through a haemolysed blood sample onto a detector, which detects the absorbance at each wavelength. Using the principles of transmission oximetry described above, the relative concentrations are determined.

Questions

What is a pulse oximeter and what does it measure?

What wavelengths of light does the pulse oximeter use?

Draw the absorbance spectra of oxyhaemoglobin and deoxyhaemoglobin indicating the isobestic points.

Describe Beer's law and Lambert's law.

What is transmission oximetry?

What is reflectance oximetry?

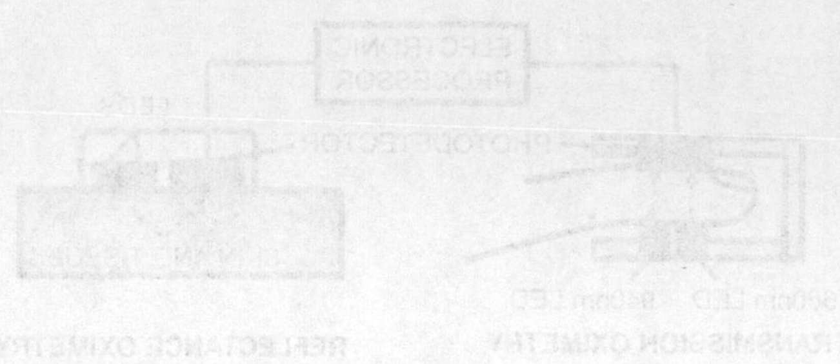

Chapter 32 Gas analysis

Basic science

Monitoring of inhaled volatile anaesthetic agents, carbon dioxide and nitrous oxide can be achieved by mass spectrometry, infrared spectroscopy, Raman scattering and photo-acoustic spectroscopy. Infrared analysers are the most commonly used in theatre.

Mass spectrometry has high accuracy and can be used to identify and measure the concentrations of carbon dioxide, nitrous oxide, oxygen and all volatile anaesthetic agents.

The mass spectrometer analyser continuously draws sample gas into a chamber where it is vaporised. A very small amount is allowed to enter a vacuum ionisation chamber via a molecular inlet leak. Here, the gas molecules lose an electron to become charged ions as an electron beam bombards them. Subsequently, these charged ions are separated according to their mass to charge ratio by either of the following techniques.

- A **sector mass spectrometer** uses a magnetic field to separate ions. Charged ions from the vacuum ionisation chamber are directed through a slit and accelerated towards a magnetic field. As the ion beam passes the magnetic field, ions with different charge and mass are sorted by varying amounts of deflection. Small and highly charged ions are deflected more than larger and less charged ions. This way, multiple beams are created that are organised according to their mass to charge ratio. The individual beams are detected by a series of plates that produce an electrical output relating to the quantity of specific ions in the sample.
- A **quadrupole mass spectrometer** consists of four charged cylindrical rods set parallel to each other and arranged around the line of accelerated particles. An oscillating electric field applied to the rods causes the ions to resonate. Only ions of a specific mass to charge ratio will have a stable flight path, allowing them to pass through the rods. All other ions will be deflected and collide with the rods and they will not be detected. Altering the strength of the electric field applied to the rods will allow detection of different ions.

When laser (monochromatic) light strikes gas molecules, most of the light will be scattered in an elastic manner (**Rayleigh scattering**). This is where there is no change in the kinetic energy of the incident photons when they scatter. Some light scattering occurs in an inelastic manner, where scattered photons gain or lose kinetic energy. This is **Raman scattering**. Scattered photons that gain energy have a shorter wavelength and higher frequency (blue shift) than the incident photon. Energy can also be lost, resulting in a longer wavelength and lower frequency (red shift).

The Raman spectrometer is fast and can be used to identify oxygen, carbon dioxide, nitrogen, nitrous oxide and all volatile anaesthetic agents. It consists of a laser beam, sample chamber, series of narrowband filters, photo-detector and a signal processing unit. Laser light (usually argon laser) is passed through a sample chamber containing the gas to be analysed. The scattered light then passes through specialised optics and a series of narrowband filters to the photo-detector. The concentration of the sample gas is proportional to the quantity of shifted photons (Raman scattered photons).

Infrared spectroscopy functions on the principle that gas molecules with dissimilar atoms absorb radiation at different frequencies from the infrared region of the electromagnetic spectrum. The absorption profile is closely related to the molecular structure of the gas. This allows identification of individual gases by analysis of the frequencies absorbed. Infrared spectroscopy can detect and measure carbon dioxide, nitrous oxide and all anaesthetic vapours. Oxygen and nitrogen are not measured because they do not contain dissimilar atoms. The amount of radiation absorbed is directly proportional to the concentration present within the sample.

Infrared analysers are dispersive or non-dispersive. **Non-dispersive** analysers are commonly used in the clinical setting. The unit consists of an infrared source, a sample chamber, a reference chamber (filled with an optically inert gas), a rotating wheel and a detector. The sample gas is passed through the sample chamber. The infrared light passes through two chambers (sample and reference) and falls on to the photosensor, which measures the intensity of the wavelength. The rotating wheel has several different optical filters that narrow the bandwidth of frequencies to best suit the profile of the gas being studied. As the wheel rotates, different gases can be analysed.

Applied science

Describe photo-acoustic spectroscopy and its advantages over infrared spectroscopy

Photo-acoustic spectroscopy, like other infrared analysis techniques, is based on the ability of gases (with dissimilar atoms) to absorb infrared light; however, the measurement methods differ. Light is shone through a rotating chopper wheel, which has the effect of rapidly switching the light on and off. The light then passes through an optical filter to enter a measurement chamber, where gas absorbs the frequency of light matching its absorption band. As the gas

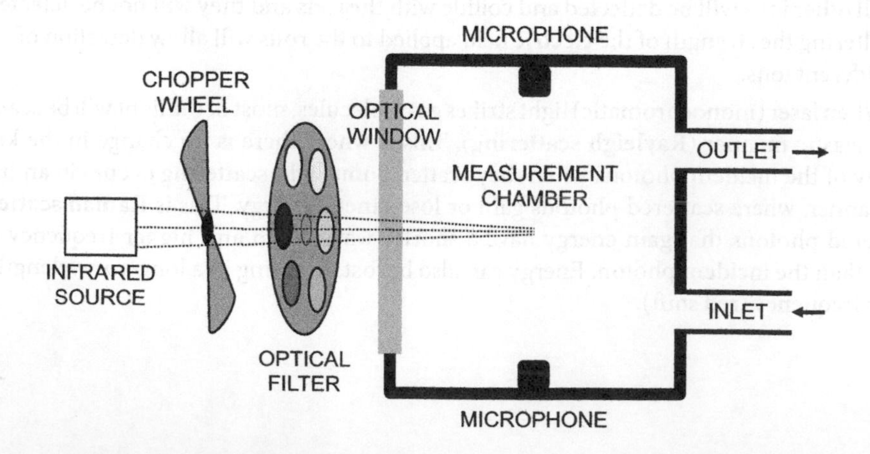

absorbs energy, its molecules gain kinetic energy, resulting in more random collisions. This causes the pressure in the measuring chamber to increase. Subsequent cessation of light supplied to the measuring chamber results in a fall in pressure. The chopped nature of the light causes the pressure to rapidly rise and fall, creating a pressure wave (sound wave). This acoustic signal is detected by two microphones, amplified and processed.

Photo-acoustic spectrometry has a higher accuracy, better reliability and does not need as frequent calibration compared to infrared absorption spectrometry.

What are the common causes of error with infrared analysers?

Sources of error with infrared analysers include the following.

- Water vapour falsely increases carbon dioxide and volatile anaesthetic readings, as it absorbs infrared at several wavelengths.
- Oxygen broadens the carbon dioxide absorption spectra, resulting in lower readings.
- Carbon dioxide and nitrous oxide have similar infrared absorption bands. Carbon dioxide absorbs strongly between wavelengths of 4.2 and 4.4 μm and nitrous oxide absorbs strongly between 4.4 and 4.6 μm. The close proximity of these bands may result in overlap of absorption peaks. Use of a narrow-band filter at 4.3 μm (the wavelength at which carbon dioxide is most strongly absorbed) can overcome this problem.
- Nitrous oxide in a gas mixture containing carbon dioxide can produce **collision broadening**. This occurs when infrared energy absorbed by a carbon dioxide molecule is emitted as a photon if it randomly collides with a nitrous oxide molecule. This enables the carbon dioxide molecule to absorb more infrared light energy, falsely increasing the measured concentration of carbon dioxide. This effect also broadens the absorption spectrum of nitrous oxide (as it is in effect absorbing energy within the spectrum of carbon dioxide). Modern analysers can compensate by measuring the concentrations of interfering agents.
- Alcohols can also cause inaccuracies by falsely elevating volatile anaesthetic agent readings.

Questions

What gases are measured in anaesthesia?

Describe the methods of anaesthetic gas analysis.

Why is infrared spectroscopy unsuitable for oxygen?

What is the collision broadening effect?

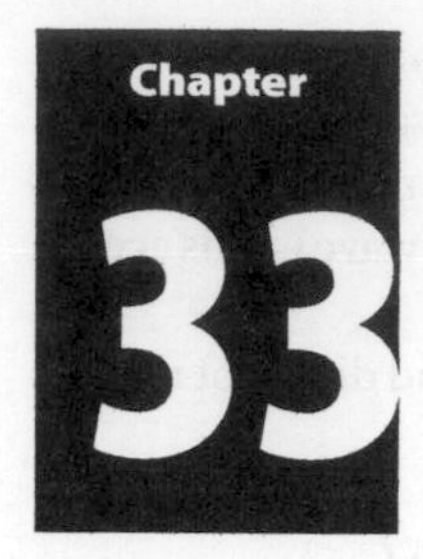

Gas analysis – oxygen

Basic science

Analysers that measure oxygen concentrations in gas mixtures are regarded as essential monitors during anaesthesia. They can be either **electrochemical** or **paramagnetic.**

Electrochemical analysers include the fuel (galvanic) cell and the Clark (polarographic) electrode. The response time of these analysers is dependent on the rate of diffusion of oxygen across a semi-permeable membrane. As a result, they are usually too slow to measure end-tidal oxygen concentrations.

The **fuel cell** consists of a **gold cathode** and a **lead anode** suspended in a **potassium chloride** solution. A membrane impermeable to ions and electrolytes but permeable to oxygen separates the cell from the sample gas. The reduction of oxygen at the cathode generates an electric current. The amount of current is proportional to the partial pressure of oxygen present in sample gas.

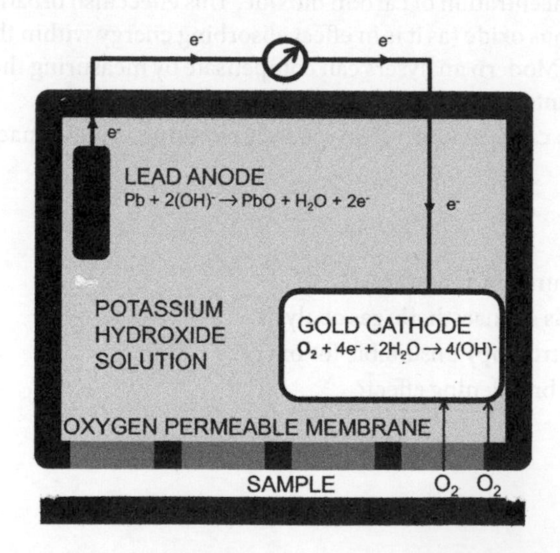

Oxygen molecules diffuse across the membrane from the sample gas into the sensor, and are reduced to hydroxyl ions at the cathode. The hydroxyl ions then migrate to and oxidise the lead anode to form lead oxide. These reactions create a flow of electrons, which can be measured. The lead anode is slowly consumed by the oxidation reactions, which limit the lifespan of the fuel cell.

The **polarographic (Clark electrode)** analyser functions in a similar manner to the fuel cell. It consists of a **noble cathode** (usually platinum) and a **silver/silver chloride anode** contained within a potassium chloride electrolyte solution. In contrast to the fuel cell, an external voltage of 0.6 V is required to cause the reduction of oxygen at the cathode. A Teflon membrane separates the cell from the gas sample.

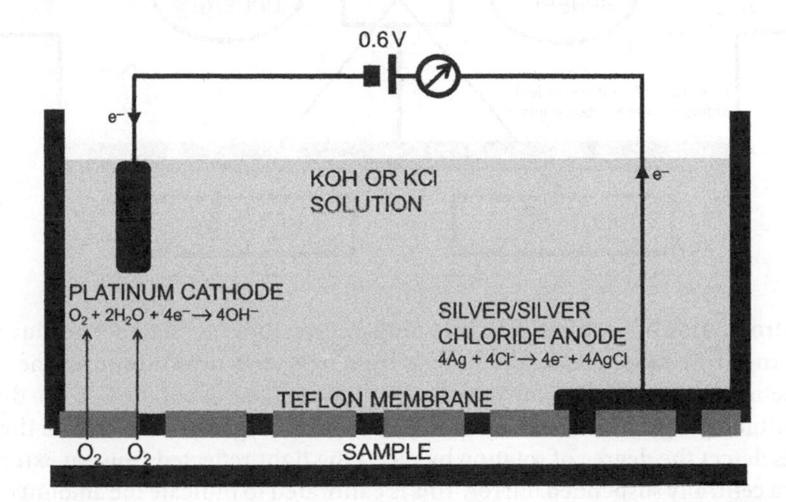

Oxygen diffuses across the Teflon membrane and the electrolyte solution to contact the polarised platinum cathode, where it is reduced to hydroxyl ions. At the silver/silver chloride anode, the silver is oxidised to silver chloride in a reaction that releases electrons.

As the chemical reactions occur, an electric current that is directly proportional to the partial pressure of oxygen will flow between the anode and the cathode. An amplifier can be used to increase the signal output.

The Clark electrode has a faster response time than the fuel cell because of its external power supply. The solubility of oxygen, similar to all gases, increases as temperature decreases. Thermistors are sometimes used in these devices to compensate for the varying amounts of dissolved oxygen at different temperatures.

Applied science

What is a paramagnetic gas? How can this property be used to measure the gas concentration in a mixture?

Paramagnetic gases are attracted to a magnetic field. Nitric oxide and oxygen are paramagnetic gases because they have unpaired electrons in their outer shells.

The original paramagnetic oxygen analysers comprise a gas sample chamber with two nitrogen-filled glass spheres in a dumbbell arrangement. The dumbbell is mounted on a rotating suspension within a strong magnetic field. The dumbbell is kept in balance by the magnetic field because nitrogen is diamagnetic, meaning it is both attracted and repelled by the magnetic field.

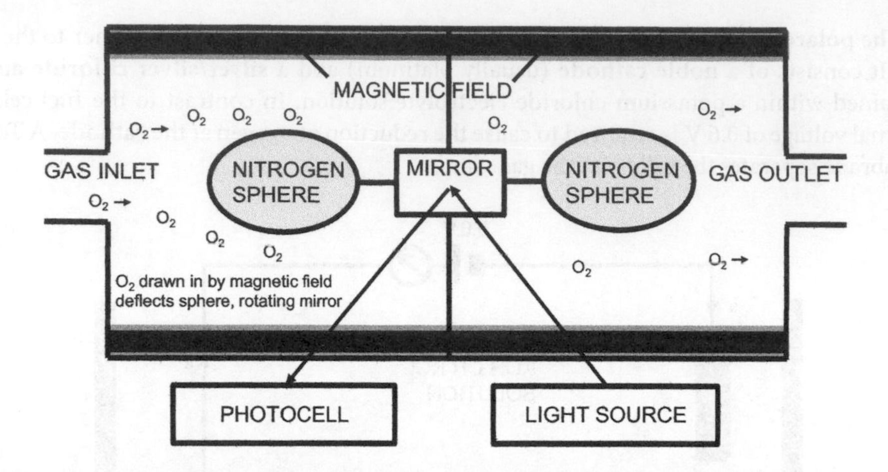

The introduction of a sample gas containing oxygen into the chamber produces a force on the nitrogen-filled spheres as the surrounding oxygen is drawn towards the strongest part of the magnetic field. The force created by oxygen causes rotation of the dumbbells by a magnitude proportional to the partial pressure of oxygen present within the sample. Photocells detect the degree of rotation by receiving light reflected from an external light source by a centrally suspended mirror. This is calibrated to indicate the amount of oxygen present.

These analysers are very accurate. However, they are affected by vibration, high flow rates and water vapour. They have a relatively slow response time.

How do newer paramagnetic analysers differ from older designs?

Newer paramagnetic analysers have overcome the disadvantages of older models by using a different application of the paramagnetic property of oxygen. Pressure variations are created by the agitation of oxygen molecules in a gas sample as it passes through a fluctuating magnetic field.

In these analysers, capillary tubes continuously deliver the sample containing oxygen and a reference gas. A pressure transducer separates the two gases where the sample lines meet. A fluctuating magnetic field at the junction of the two capillary tubes creates a pressure differential between the two gas streams, which is detected by the pressure transducer. The pressure difference across the transducer is proportional to the difference in the oxygen partial pressure in the sample compared to the reference gas. The output from the transducer is an electrical signal that is used to display the amount of oxygen present.

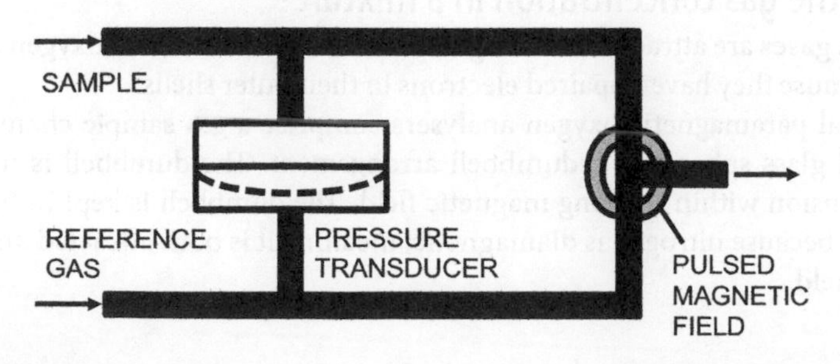

Questions

How can oxygen in gas mixtures be measured?
Describe the fuel cell and the Clark electrode.
Why does the Clark cell require an external power source?
What is the purpose of the thermistor in electrochemical sensors?
How does the paramagnetic oxygen analyser work?
What other paramagnetic gases do you know of?

Chapter

34

CO_2 analysis

Basic science

Direct methods of CO_2 analysis include the following.

Severinghaus electrode: This is a pH electrode, modified with a covering rubber membrane containing sodium bicarbonate solution. The membrane is permeable to CO_2, which diffuses into the bicarbonate layer.

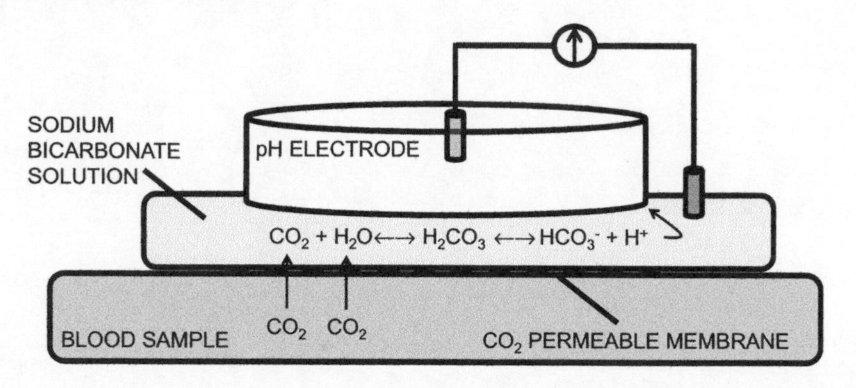

The pH of the bicarbonate solution falls in proportion to the CO_2 concentration, which is measured by the adjacent pH electrode. The Severinghaus electrode has a response time of 2–3 minutes, and needs frequent calibration and temperature maintenance at 37°C. Samples must be free of air bubbles to avoid CO_2 diffusing out of solution, and analysis should be undertaken immediately to prevent ongoing red cell metabolic activity increasing the CO_2 content.

Cutaneous electrode: A Severinghaus electrode is placed directly upon skin heated to 42–43°C. The heating causes capillary dilatation, enabling CO_2 to diffuse out through the skin and into the measuring electrode. Cutaneous electrodes are inaccurate in low cardiac output states and on oedematous skin, and prolonged use can cause burns.

Intravascular electrode: These comprise a needle probe made of CO_2-permeable silicon, containing a CO_2-sensitive fluorescent dye such as 1-hydroxypyrene-3,6,8-trisulfonate. Fluorescence changes linearly with the CO_2 concentration, which is measured by a fibre-optic light source and photo-detector. This gives a real-time measure of CO_2 concentration, but is invasive and relatively expensive.

Indirect methods of CO_2 analysis include the following.

Calorimetric devices: These comprise an encased pH-sensitive indicator placed onto the endotracheal tube. Expired CO_2 creates a visible colour change in the indicator, but no quantitative measure of CO_2.

Capnograph: This provides a continuous measurement and display of expired CO_2 concentration. It is an essential anaesthetic monitor, which uses infrared spectroscopy. The capnograph underestimates the arterial pCO_2 by 0.5–0.8 kPa, although this can be greater in cases of respiratory disease with ventilation perfusion mismatching, or low cardiac output states. They can be either sidestream or mainstream.

Sidestream capnographs continuously aspirate 150 $ml \cdot min^{-1}$ of gas from a connector at the endotracheal tube. The sample passes through a water trap, then is analysed by an infrared spectrometer before it is returned to the breathing system or scavenged.

Mainstream capnographs incorporate a spectrometer between the endotracheal tube and breathing system. The infrared beam passes directly through expired gas onto a detector on the opposite side. The spectrometer is heated to >37°C to prevent water condensation.

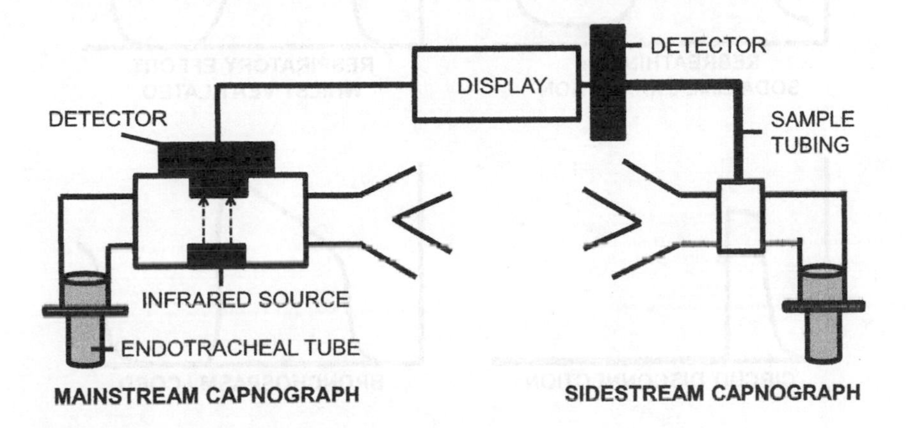

Capnometers: Similar to mainstream capnographs, but they omit the waveform display, displaying only the end tidal carbon dioxide ($etCO_2$) concentration value.

Additionally, mass spectrometry, gas chromatography, Raman spectrometry and the Siggaard–Andersen nomogram can also be used.

Applied science

What clinical information can be gained from a capnograph?

The capnograph can:

- confirm endotracheal tube placement,
- confirm ventilation,
- guide minute ventilation during positive pressure ventilation,
- determine respiratory rate in spontaneously breathing patients.

Waveform analysis can also yield useful information. The normal capnograph consists of four phases.

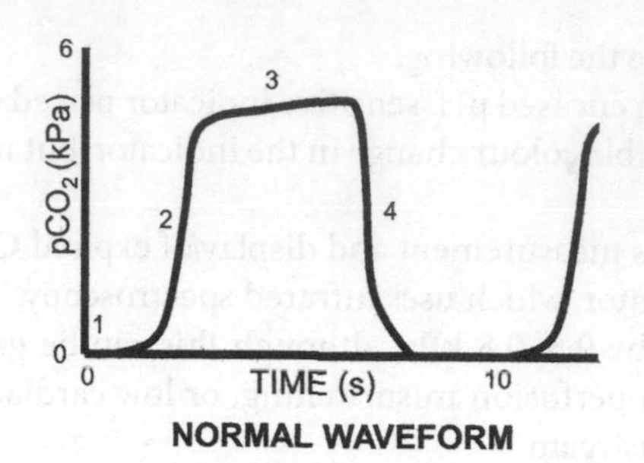

NORMAL WAVEFORM

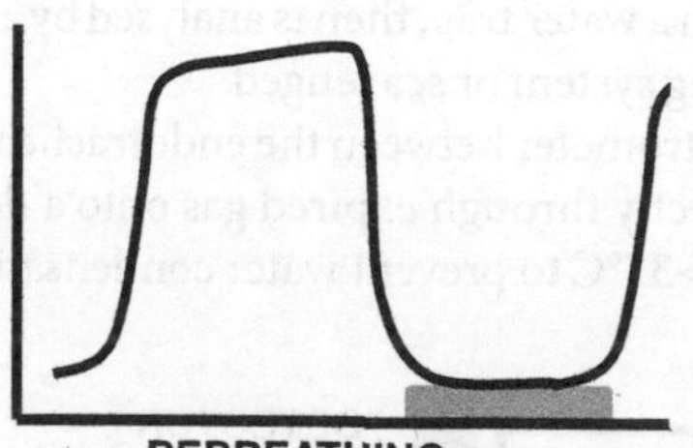

REBREATHING or SODA LIME EXHAUSTION

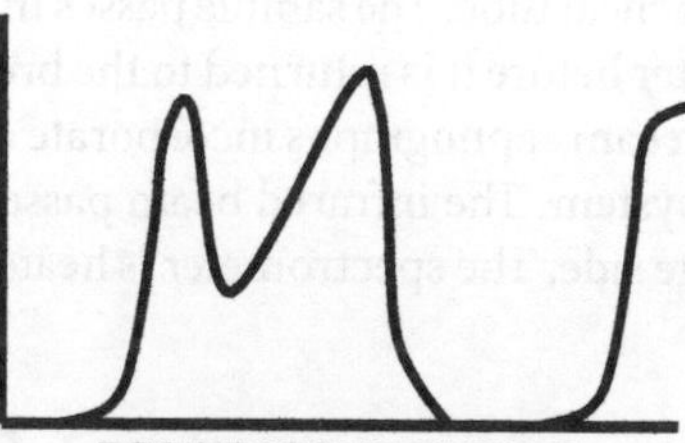

RESPIRATORY EFFORT WHILST VENTILATED

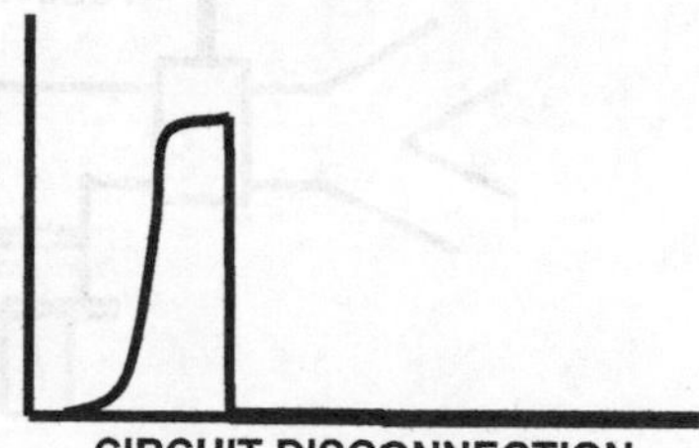

CIRCUIT DISCONNECTION

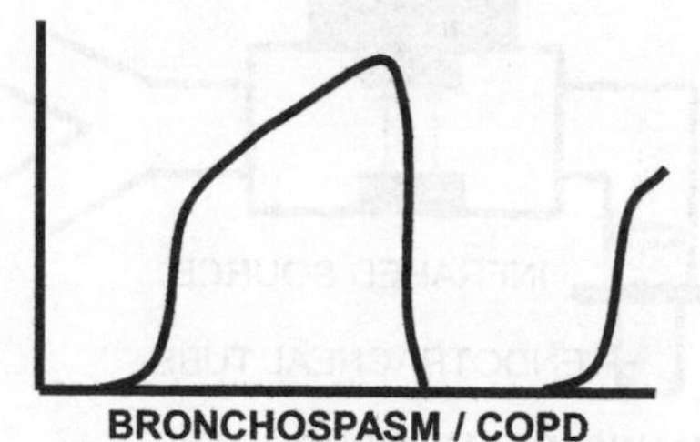

BRONCHOSPASM / COPD

DECREASING $ETCO_2$ (PULMONARY EMBOLUS / DIMINISHING CARDIAC OUTPUT)

CARDIOPULMONARY RESUSCITATION WITH RETURN OF SPONTANEOUS CIRCULATION

What are the advantages and disadvantages of the sidestream and mainstream capnographs?

Sidestream advantages

- Light-weight connections.
- Can be used in uncommon positions, e.g. prone.
- Does not require sterilisation.
- Can be used with non-intubated patients (e.g. sampling from a Hudson mask to monitor respiratory rate during sedation).

Sidestream disadvantages

- Delayed response time as sample travels from endotracheal tube to infrared unit.
- Sample tube kinking/obstruction is common.
- Loss of anaesthetic gases (if sample is scavenged).
- Water vapour invalidates result, so needs a water trap/filter.
- The Bernoulli effect introduces pressure drops across the sampling tube, which alters the sample partial pressure of CO_2.
- Dispersion of gas within sampling tube may distort the waveform.

Mainstream advantages

- Portable.
- Faster response time.
- Avoids sampling tube and return tube/scavenging problems.

Mainstream disadvantages

- Increased apparatus dead space (due to the connector).
- Bulky and adds weight to the airway.
- Use is limited to patients with endotracheal/supraglottic airways.
- Risk of facial burns from heating element.
- May become clogged with secretions.

Questions

Describe some of the direct and indirect methods of CO_2 analysis.
What is the difference between a capnometer and a capnograph?
How do mainstream and sidestream capnographs differ?
Describe the function of the Severinghaus electrode.

Chapter 35

Blood gas analysis

Basic science

An **acid** is a substance that can donate a proton (H^+) or accept an electron pair in reactions. A **base** accepts protons or donates an electron pair. The pH scale measures hydrogen ion concentration:

$$pH = -\log_{10}[H^+]$$

The logarithmic scale produces a 10-fold change in hydrogen ion concentration per unit change in pH. Modern blood gas machines rapidly **measure** several variables including pH, pCO_2, PO_2, oxyhaemoglobin concentration and ions (Na^+, K^+, Cl^- and Ca^{2+}). They **calculate** some variables including actual bicarbonate, standard bicarbonate and base excess.

The **pH electrode** measures pH. It consists of a silver measuring electrode which is coated with silver chloride and enclosed in a buffer solution of HCl which maintains the pH at 7. The electrode tip is composed of specialised H^+ ion-sensitive glass. When placed in a solution the glass softens, forming a semi-solid gel membrane. Hydrogen ions diffuse into the outer layer of the glass, displacing ions such as lithium and sodium toward the inner layers. This creates a positively charged glass surface, which attracts anions (Cl^-) from the buffer solution. An ion gradient is established within the glass electrode; this is dependent on the H^+ concentration in the sample. Thus a potential difference is created. The potential difference is measured against a reference electrode that is connected by a lead wire and placed into the same sample being measured. The silver/silver chloride reference electrode is contained in a saturated 3 mol l^{-1} KCl solution, which gives it a fixed potential. It is encased in a non-permeable shell with a diaphragm at its tip, allowing continuity with the sample. Calomel electrodes were used previously, but are now avoided due to the toxicity of mercury.

Modifying the properties of the measuring membrane can make it specific to other ions including Na^+, K^+, Cl^- and Ca^{2+}. In this way, their concentrations can be measured using similar apparatus. Membranes used include the following.

- Glass: framework of silicate with interstitial sites for H^+.
- Crystal: lattice containing defined gaps for ions of all types.
- Polymers: contain an ionophore (molecule) that specifically binds the ion to be measured.

The pH electrode is shown below.

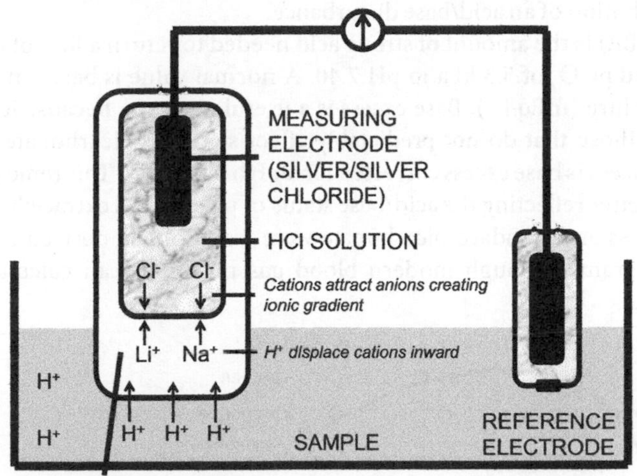

Electrodes are prone to certain errors.

- Temperature affects the dissociation of H^+ ions from bicarbonate. Therefore, the apparatus must be kept at 37°C.
- Blockage of ion-sensitive channels or reference electrode diaphragm by precipitated silver chloride prevents functioning. Regular cleaning and maintenance are recommended.
- Damage or leakage from the electrode into the sample negates potentials developed, requiring electrode replacement.
- Ion selectivity of a membrane is not 100% specific. The presence of other ions in high concentrations can influence the results. **Standard addition** can be used to overcome this, whereby defined volumes of the ion to be measured are added to the sample in steps, and the effect measured. The initial concentration can then be calculated by extrapolation.
- Drift affects the precision and accuracy over time, and can be prevented by regular 2-point calibration with buffer solutions of known pH.

Applied science

How are the bicarbonate concentration and base excess calculated?

Actual bicarbonate is the true concentration of HCO_3^- in a sample. It is derived from the **Henderson Hasselbalch equation** by inserting the values of pH and pCO_2 measured by the blood gas machine:

$$pH = pKa + \log\frac{HCO_3^-}{CO_2}$$

Standard bicarbonate is the concentration of plasma bicarbonate under standardised conditions: temperature 37°C, pCO_2 of 5.3 kPa and fully saturated with oxygen. By keeping

pCO_2 constant, the respiratory component is eliminated, resulting in a figure that reflects the metabolic contribution of an acid/base disturbance.

Base excess (BE) is the amount of strong acid needed to return a litre of fully oxygenated blood at 37°C and pCO_2 of 5.3 kPa to pH 7.40. A normal value is between −2 and +2 milliequivalents per litre ($mEq{\cdot}l^{-1}$). Base excess is a useful measure, because it accounts for all acids, including those that do not predictably affect standard bicarbonate concentrations. **Standard base excess** is base excess corrected for an Hb of 5 $g{\cdot}dl^{-1}$. This removes the buffering effect of blood, better reflecting the acid/base status of plasma and extracellular fluid.

The base excess and standard bicarbonate were traditionally derived from a Siggaard-Andersen nomogram, although modern blood gas machines can calculate values using in-built formulae.

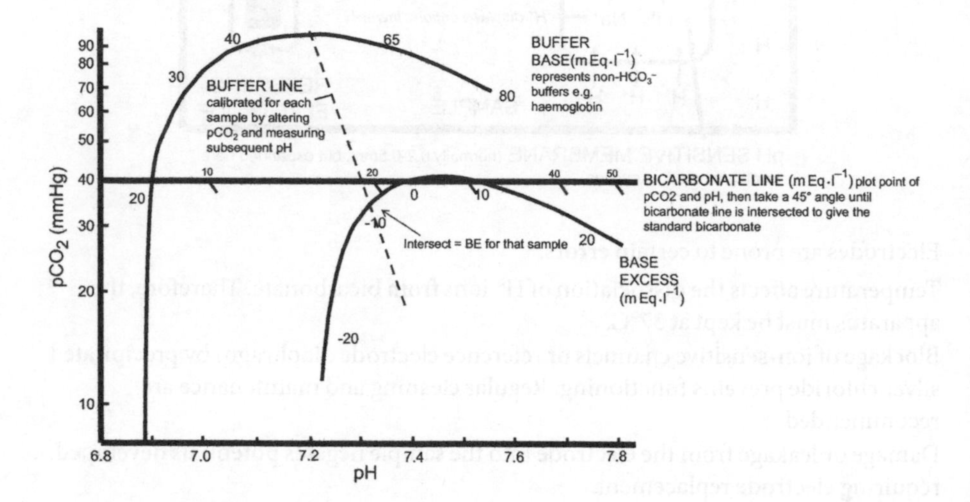

Describe pH stat and alpha stat and their clinical relevance

Homeostatic mechanisms maintain plasma pH at 7.4 and pCO_2 at 5.3 kPa. When therapeutic hypothermia is used, such as in cardiopulmonary bypass, controversy exists whether these 'normal' values should apply.

With a fall in temperature, CO_2 solubility increases so partial pressure decreases, causing pH to rise. To return pH to 7.4 and pCO_2 to 5.3 kPa, CO_2 needs to be added, increasing the total body content of carbon dioxide. This is the principle underlying pH stat. Because blood gas samples are conventionally warmed to 37°C prior to analysis, temperature correction is required to account for the initial sample temperature. This can be obtained from nomograms.

Alpha stat proponents advocate that it is more important to maintain **intracellular electrochemical neutrality**, rather than pH. In neutral states, the intracellular $[H^+]$:$[OH^-]$ ratio is 1:1, with water acting as the primary source of hydrogen ions. The dissociation constant of water (pKw) rises as temperature falls, which leads to a corresponding fall in $[H^+]$. Hydroxyl ions fall in tandem with $[H^+]$, as they too are influenced by the pKw. This maintains the electrochemical neutrality of the cell despite pH rising, as pH is simply a measure of H^+ ion concentration.

With pH stat, the rise in total body CO_2, shifts the intracellular ratio of $[H^+]:[OH^-]$. This subsequently alters the fraction of unprotonated imidazole groups of histidine (known as 'alpha'), which are the predominant intracellular buffer. This affects the structure and function of cellular enzymes and proteins. Alpha stat aims to maintain alpha constant, by avoiding correction for temperature.

Neither technique has been shown to have superior outcomes. Cerebral blood flow is greater with pH stat due to higher pCO_2, but it may accelerate endothelial damage and increase intracerebral pressure. Alpha stat may result in faster recovery of the ventricles.

Questions

How is pH measured?

Which blood gas variables are measured? Which are calculated?

What is the significance of the Siggaard–Andersen nomogram?

Section 8

Suction, CO_2 absorption and cleaning

Chapter 36

Suction and vacuum

Basic science

Vacuum is defined as a space entirely devoid of matter. This is a space with a negative pressure with respect to atmospheric pressure.

Suction is the application of a vacuum to cause the movement of solids, liquids and gases. Gauge pressures, commonly in mmHg or cmH_2O, measure the force of vacuum suction produced. This is the vacuum force required to raise a column of water or mercury a specified distance. Vacuum **regulators** attached to suction equipment adjust the force of the suction applied to the patient.

Most hospitals supply vacuum via a pipeline system, which is generated by a **vacuum pump**. Portable suction devices are either vacuum pumps or venturi systems.

The pipeline vacuum consists of pumps, a receiver or reservoir, vacuum sensors, valves and pipes. Two or more pumps remove air from the receiver, creating a **negative or sub-atmospheric pressure** within, which acts as the driving force for suction. An opening in the receiver creates a pressure gradient between the atmosphere and the receiver, allowing gas to flow into the receiver. The flow rate is dependent on the pressure gradient and flow continues until the pressure difference is equalised. Continual pump function maintains the pressure gradient and suction.

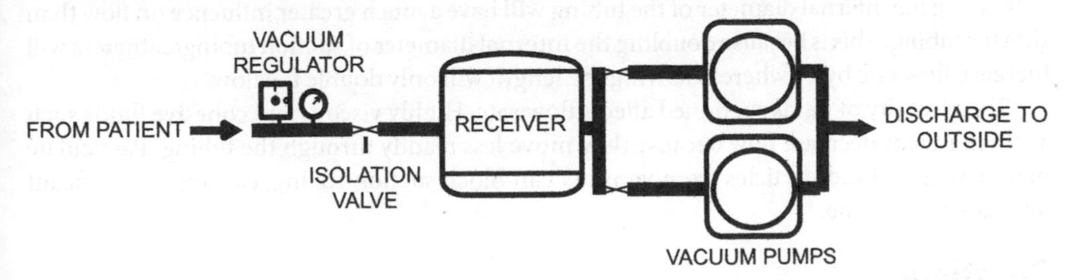

Venturi systems utilise the Bernoulli principle to create a negative pressure. They are driven by compressed gas, but are limited by weak flow rates and loud noise.

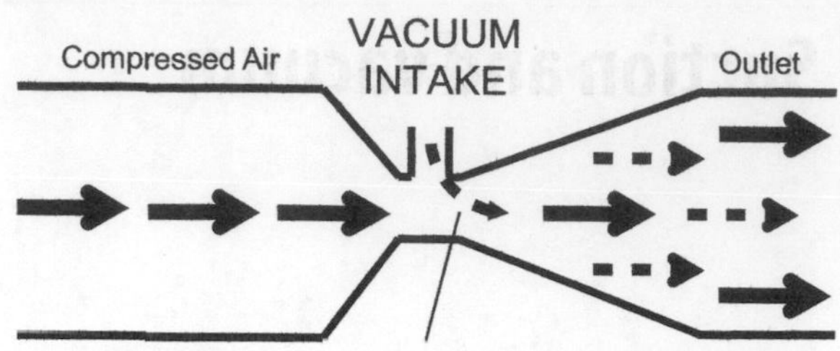

Applied science

What factors affect the flow rate of a suction system?

The flow rate of suction equipment refers to the amount of fluid moved through the unit per unit time. This is influenced by the **pressure gradient** between the regulator and suction catheter, the **resistance** of the equipment and the **viscosity** of the fluid.

Vacuum regulators vary the maximum amount of negative pressure supplied from the vacuum source to the patient end of the suction apparatus (collection canister and suction tubing). They often comprise an adjustable spring/diaphragm to set and maintain a constant vacuum level. Adjusting the regulator to increase the pressure gradient from the vacuum source to the patient end will increase the flow rate through the system.

The total resistance of the system defines the maximum achievable flow rate. The regulator, collection canister, tubing and connectors contribute to the resistance in the system. The most notable factor increasing resistance and thus reducing flow rate is the suction tubing attached to the collection canister. Ideally, this tubing should be short and have the largest internal diameter possible. When flow is laminar, Hagen Poiseuille's law will apply. Therefore, increasing the internal diameter of the tubing will have a much greater influence on flow than shorter tubing. This is because doubling the internal diameter of suction tubing/catheters will increase flow rate by 2^4, whereas halving the length will only double the flow.

The viscosity of fluids suctioned affects flow rate. Highly viscous and cohesive fluids such as thick mucus decrease flow because they move less readily through the tubing. Particulate matter such as food particles from vomitus can block suction tubing, causing a significant decrease in flow rate.

Questions

What is meant by the terms suction and vacuum?
How is the vacuum for suction equipment generated?
What are the components of a pipeline suction system?
What is a suction/vacuum regulator?
How does the resistance of the equipment affect flow?

Chapter 37

CO_2 absorption

Basic science

Soda lime and **baralyme** are the two most commonly used agents for carbon dioxide absorption in anaesthesia.

Baralyme contains approximately 80% calcium hydroxide and 20% barium hydroxide, with or without 1% potassium hydroxide. Soda lime consists of calcium hydroxide, sodium hydroxide and potassium hydroxide in variable amounts (depending on the manufacturer).

Soda lime constituent	Percentage (%)
Calcium hydroxide	80–94
Sodium hydroxide	4–5
Potassium hydroxide	1

Water is required in order for the chemical reactions to occur with both soda lime and baralyme. The optimal water content for carbon dioxide absorption is between 14% and 19%.

Silica is added to form calcium and sodium silicate. This makes the soda lime harder and reduces the formation of dust. The efficacy of carbon dioxide absorption will vary inversely with hardness.

The granules are size **4–8 mesh.** This refers to the granule size that will pass through a mesh with 4–8 strands per square inch.

Colour indicators that are pH-sensitive are added to allow a visual assessment of the integrity of soda lime and baralyme. As the absorbent capacity is exhausted, CO_2 causes the pH to fall. This changes the colour of the pH indicator. The colour change is pH indicator-specific (e.g. ethyl violet changes from white to purple).

Complications that can occur with the use of soda lime and baralyme include the production of **compound A,** and the accumulation of **carbon monoxide** (CO). Compound A is produced as a degradation product when sevoflurane reacts with sodium hydroxide or potassium hydroxide. It has been shown to cause transient nephrotoxicity in rats. Carbon monoxide is produced by the reaction of halogenated vapours with the CO_2 absorbers. The reaction is more pronounced when the absorber is dry, at higher temperatures, and occurs to a greater extent when desflurane reacts with baralyme.

Applied science

Describe the reactions that occur in soda lime

$$1.\ CO_2 + H_2O \rightleftharpoons H_2CO_3$$

Carbonic acid is formed by the reaction between carbon dioxide and water.

$$2.\ H_2CO_3 + 2NaOH \rightarrow Na_2CO_3 + 2H_2O + \text{Heat}$$

Carbonic acid reacts with sodium hydroxide to form sodium carbonate. This step also produces water and heat.

$$3.\ Na_2CO_3 + Ca(OH)_2 \rightarrow CaCO_3 + 2NaOH$$

The final set of reactions between sodium carbonate and calcium hydroxide create calcium carbonate and regenerate sodium hydroxide.

$$4.\ H_2CO_3 + Ca(OH)_2 \rightarrow CaCO_3 + 2H_2O$$

The fourth reaction can occur, but does so at a very slow rate. It is the direct reaction between carbonic acid and calcium hydroxide to form calcium carbonate.

In reactions 2 and 3 described above, potassium hydroxide and potassium carbonate can be substituted for sodium hydroxide and sodium carbonate, respectively, to describe the process that occurs with potassium hydroxide.

Questions

What CO_2 absorbers are commonly used?
What are the complications that can occur with CO_2 absorbers?
What is the chemical composition of soda lime?
What size are the granules?
How can you tell when a CO_2 absorber has been exhausted?
Give an example of a colour indicator used.

Chapter 38

Cleaning, disinfection and sterilisation

Basic science

Cleaning is the physical removal of contaminants from an object by mechanical or manual action such as washing. A cleaned object is not free of microorganisms and can still transmit infections. Cleaning precedes disinfection and sterilisation.

Disinfection reduces the number of microorganisms on objects, but does not destroy bacterial spores. High-level disinfection kills microorganisms such as *Mycoplasma* and viruses. Low-level disinfection typically does not kill *Mycoplasma*. The disinfection process can involve the following.

- **Steam**: the process involves exposing the equipment to dry saturated steam at a temperature of 73°C at atmospheric pressure for a minimum of 10 minutes. The low-temperature steam kills most vegetative microorganisms and some heat-sensitive viruses. This process can be used for breathing tubes, bite blocks, face masks and laryngoscope blades. The main advantages are that it is non toxic, and less damaging to equipment. Its main disadvantage is that there is an increased risk of contamination because the equipment is wet.
- **Pasteurisation** (hot water disinfection): this process involves immersing an object in boiling water for a specified time and temperature, for example, 70°C for 30 minutes. Boiling kills most non-spore-forming microorganisms, viruses and some heat-sensitive bacterial spores. The main advantage is that it is less damaging to equipment than autoclaving, and is non-toxic. However, heat may cause damage to some items.
- **Liquid chemicals**: alcohols (ethanol or isopropanol), aldehydes (formaldehyde and glutaraldehyde) and oxidising agents (hydrogen peroxide) are used. The equipment must be cleaned beforehand to ensure good contact. It is then immersed in the liquid chemical for a specified period of time. The immersion time is dependent upon the chemical used and the level of disinfection required. Chemicals are advantageous because they have a broad spectrum disinfection range. However, they are potentially toxic, volatile and can cause damage to equipment.
- **Sterilisation** is used for the complete destruction or removal of all microbial organisms, including viruses and bacterial spores, fungi and protozoa. The sterilisation process can involve the following.
- **Autoclaves:** saturated steam under pressure is used to achieve complete destruction of all microorganisms. This requires direct contact between the equipment and the saturated steam, complete removal of air, and precise control of time, temperature and pressure.

It is relatively fast, highly effective and non-toxic. However, it is unsuitable for heat and pressure sensitive equipment, such as fibre-optic endoscopes, and may cause corrosion of metal surfaces.

- **Dry air ovens**: dry heat can be used to sterilise non-aqueous liquids, semi-solids (glycerine, oils and waxes), powders, needles and glass syringes. Typical times and temperatures required are 160°C for 2 hours, 170°C for 1 hour, and 180°C for 30 minutes. This process is neither toxic nor corrosive, but it is more time-consuming compared to the autoclave.
- **Ethylene oxide**: gas sterilisation with ethylene oxide is usually carried out at 20°C to 60°C for 2–24 hours. Ethylene oxide can be used with or without a diluent gas (nitrogen or carbon dioxide). Sterilised items need to be aerated prior to use. Ethylene oxide is toxic and is unsuitable for use on ventilatory or respiratory equipment.
- **Gas plasma sterilisation**: this uses a cloud of ionised gas containing reactive ions. Applying energy to certain gases such as hydrogen peroxide forms an ionised gas. The reactive ions interfere with cellular metabolism to cause sterilisation.
- **Radiation sterilisation**: gamma radiation is used to sterilise equipment. The dose and time depend on the thickness and volume of the equipment. It requires expensive equipment.
- **Chemical**: 2% glutaraldehyde can be used, but prolonged immersion times are required to achieve full sterilisation.

Applied science

Describe how fibre-optic endoscopes are processed after use

Fibre-optic endoscopes are regarded as semi-critical devices and therefore require at least a high-level disinfection. They are heat-sensitive devices that are difficult to clean. The following steps describe the cleaning process for fibre-optic endoscopes.

1. After use, the fibre-optic endoscope should be disconnected, wiped clean and the ports flushed with water.
2. Suction valves should be dismantled. These components are usually heat-stable and can often be steam sterilised.
3. The fibrescope should be inspected for damage and tested for leaks. This should occur prior to immersion in any solution because leaks could damage the fibrescope. Any faulty items should be decommissioned until they have been repaired.
4. A waterproof cap is placed over the video connector. The channels are flushed with enzymatic detergent and the fibrescope is completely submerged in the enzymatic cleaner for a specified duration.
5. External surfaces and ports are cleaned with a disposable brush to remove any debris. This should be repeated until there is no visible debris on the brush.
6. The fibrescope is rinsed with sterile water and hung vertically to allow drainage.
7. High-level disinfection occurs by submersion in a chemical disinfectant such as glutaraldehyde or hydrogen peroxide. It is important to ensure all ports of the scope are filled with disinfectant.
8. Following disinfection, the fibrescope and all channels are thoroughly rinsed with sterile water followed by a 70% to 90% alcohol solution. All channels are then purged using forced air. This reduces contamination of the endoscope by waterborne pathogens and facilitates drying.

How do sterilisation, high-, intermediate- and low-level disinfection differ? How do you select which technique to use?

Disinfection can be subdivided into low-, intermediate- and high-level disinfection. It should be noted that disinfection at any level will not remove all bacterial spores. This can only be achieved by sterilisation. All **critical** medical devices that breach patient tissues or enter the vascular system must be sterilised.

Low-level disinfection is suitable for the elimination of fungi, vegetative bacteria and most viruses (except small or non-lipid viruses). This level of disinfection is suitable for **non-critical** items (those that touch intact skin) such as blood pressure cuffs and monitoring cables. Suitable methods for low-level disinfection include chemical exposure for ≥1 minute with alcohol (ethanol or isopropanol), sodium hypochlorite (bleach), or germicidal detergent (phenolic, iodophor or quaternary ammonium) solutions.

Intermediate-level disinfection extends the disinfection spectrum to include the small viruses such as polio and rhinovirus. It can be suitable for some semi-critical items.

High-level disinfection will eliminate fungi, vegetative bacteria, small and medium viruses and mycobacteria. High-level disinfection is required for all **semi-critical** patient care devices such as endoscopes. These are items that contact non-intact skin and mucous membranes. Chemical methods for high-level disinfection require exposure times from 12 to 30 minutes. Suitable chemicals include aldehydes (formaldehyde and glutaraldehyde), oxidising agents (hydrogen peroxide) and sodium hypochlorite. Wet pasteurisation at 70°C for 30 minutes with detergent cleaning is another alternative.

Questions

How do cleaning, disinfection and sterilisation differ?
What methods exist for each of the above?
Describe some advantages and disadvantages for each.
What equipment is not suitable to be autoclaved?
What do the terms critical, semi-critical and non-critical items mean?

Section 9 Radiology

Chapter 39

MRI

Basic science

Magnetism is the force produced by the movement of charge, resulting in attractive and repulsive effects. The SI unit of magnetic flux density is the **tesla (T)**. One tesla is roughly 20 000 times the magnetic field at the Earth's surface. Another commonly used unit is the **gauss**; one tesla = 10 000 gauss. Field strength is strongest at the core of a magnet, and diminishes exponentially with distance.

In magnetic resonance imaging (MRI), patients lie in a hollow cylindrical bore containing a static magnetic field strength of between 0.5 and 3 T. This field is created by a **primary** superconducting electromagnet, maintained at –273°C by immersion in liquid helium. At this temperature friction forces are minimal, maximising efficiency.

Atoms with odd numbers of protons such as hydrogen **precess** (spin around an axis) creating a small magnetic field. When placed in the strong magnetic field of the MRI, these atoms align parallel to it, known as the **ground state.** A radiofrequency (RF) pulse created by a secondary **gradient magnet** in a perpendicular direction to the primary field causes some of these atoms to realign in an **excited** (anti-parallel) direction. The atoms then relax to realign with the primary magnetic field, returning to ground state and emitting electromagnetic energy which is detected by the MRI. The rate of relaxation of atoms is tissue-specific, allowing identification of body structures. Intravenous contrast such as **gadolinium** alters relaxation rates of hydrogen nuclei to improve differentiation between tissues.

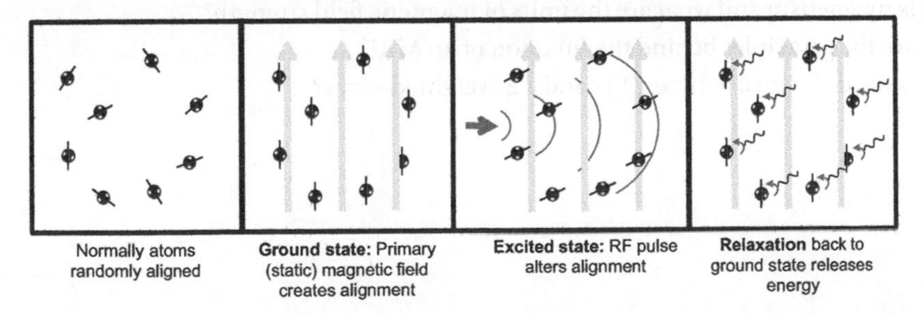

In addition to precession, atoms also have a separate rotational axis. The RF pulse causes rotation of atoms to be in phase (i.e. rotate at similar speeds). MRI scanners use this concept to perform **time weighting. T1 weighting** examines the relaxation time of atoms according to their magnetic vector alignment. **T2 weighting** examines the loss of rotational phase as atoms return to the ground state. Water appears bright on T2- but dark on T1-weighted images.

Applied science

What hazards does the MRI pose and how are they minimised?

The MRI has little direct effect, but poses several indirect hazards.

Ferromagnetic objects are attracted to magnetic fields. Gas cylinders, hypodermic needles, etc., become dangerous projectiles within the MRI. Intracranial clips may dislodge, and pacemakers and metallic heart valves may malfunction. Classification of equipment dictates their scope of use within the MRI:

- **MR safe**: non-magnetic and non-conductive equipment that is safe to use anywhere within the scanner room.
- **MR conditional**: may contain MR-sensitive components, but can be used within a set proximity of the scanner. The **5-gauss line** is the perimeter around the scanner where static magnetic field strength is >5 gauss, and is often the proximity limit for MR conditional devices. A **50-gauss line** is also present in some scanners, although less common.
- **MR unsafe**: should not be brought into the scanner room.

Electromagnetic induction in standard **monitoring** cables interferes with monitoring equipment, and may cause burns to patients. To avoid this, monitoring electrodes are made of paramagnetic material (such as nickel), and cables are encased in a **Faraday's cage**, which provides shielding. The pulse oximeter uses fibre-optic cables to transmit signals to an analyser located outside the room. The cables pass through a special hole in the wall called the **wave guide.**

Acoustic noise produced by the MRI can be up to 85 dB, which may impair hearing and masks audible alarms. Ear protection is essential for individuals entering the room while scanning. Cooling, used to maintain magnet efficiency, may accelerate heat loss from patients leading to **hypothermia.** Finally, an emergency shut-down function (**quenching**) discharges the liquid helium into the atmosphere. If an individual were exposed to this, asphyxiation and frostbite may occur.

Questions

What is magnetism and what are the units of magnetic field strength?
What are the principles behind the function of an MRI?
What is the difference between T1- and T2-weighted scans?

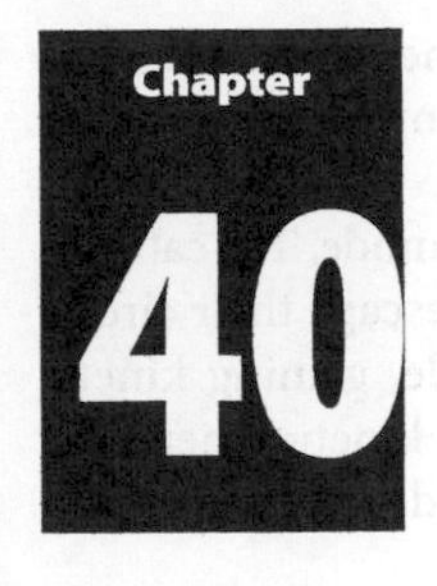

Ionising radiation

Basic science

Atoms contain a nucleus of protons and neutrons, orbited by electrons. The **atomic number** is the number of protons, which also defines an element's identity. The **atomic mass** is the sum of the protons and neutrons. **Isotopes** are atoms of the same element with differing atomic masses. Examples include:

$^{\text{atomic mass}}_{\text{atomic number}}[\text{element}]$	$^{1}_{1}H$	$^{2}_{1}H$	$^{3}_{1}H$
STANDARD NOTATION	HYDROGEN	DEUTERIUM	TRITIUM

Some isotopes, such as tritium, are unstable and undergo radioactive decay. This is a random process that follows an exponential pattern; the **half-life** of an isotope is the time taken for its mass to halve. Rates of decay are measured in **Becquerels (Bq)**.

$$1\ Becquerel = 1\ nuclear\ decay \times second^{-1}$$

The **curie** was traditionally used to measure decay, which is based on the activity seen in 1 gram of radium-226, which equals 3.7×10^{10} Bq.

Decay emits radiation, which can be of three types.

- **α-radiation**: a helium nucleus (two protons and neutrons) is emitted.
- **β-radiation**: a neutron splits into a proton and an electron, emitting the electron at high velocity. Occasionally a positively charged electron is emitted, known as a positron.
- **γ-radiation**: electromagnetic waves of wavelength $<10^{-12}$ m. They occur whenever α- or β-radiation is emitted.

These types of radiation are classed as **ionising**, as they are capable of displacing electrons from around the nucleus of atoms they collide with to form ions. Ionising radiation is potentially damaging to body cells and DNA. The SI unit for absorbed dose is the **gray (Gy)**, defined as 1 joule of energy deposited in 1 kilogram of mass.

The type of radiation and the tissues being irradiated influence the clinical outcome. Alpha radiation is 20 times more ionising than gamma radiation, and rapidly dividing tissues (such as ovarian tissue) are more susceptible to ionising effects. For this reason, the **sievert (Sv)** is a more useful clinical measure of radiation exposure; it is the amount of radiation equivalent in biological effectiveness to one gray of gamma radiation. The sievert is generally used for low levels of radiation absorption, and is tissue-specific.

X-rays are waves of electromagnetic radiation, similar to γ-rays. The main difference is their origin; γ-rays originate from the nuclei of decaying radio-isotopes, X-rays from electrons in X-ray tubes.

X-ray tubes are vacuum containers with a heated cathode and a metal anode. The cathode emits electrons by **thermionic emission**, where heat allows electrons to escape their circuit into the vacuum. They accelerate towards the positively charged anode, gaining kinetic energy. Collision with the anode causes rapid deceleration, releasing kinetic energy as **Bremsstrahlung** (German for 'braking energy'), which are X-rays. The anode can be tungsten, molybdenum or copper, all of which have high melting points.

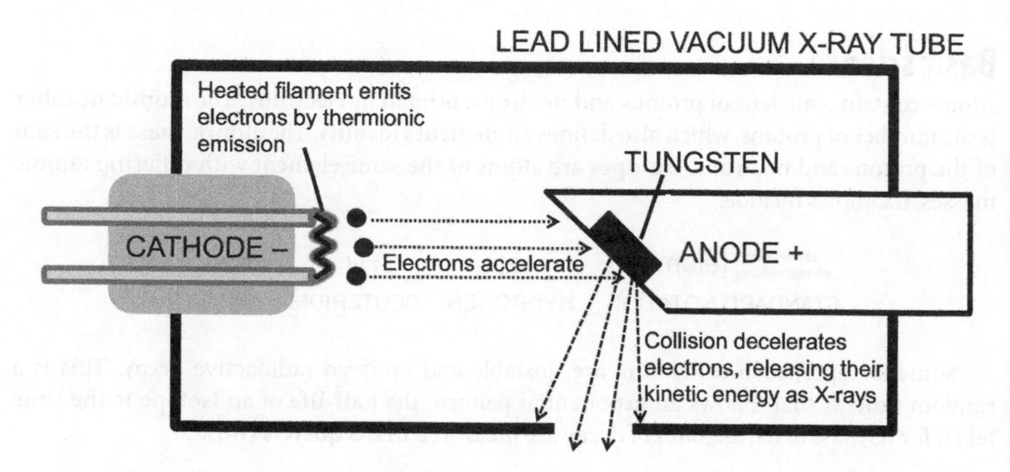

X-ray machines are rated according to voltage and current. **Voltage** is usually 30–150 kV, and determines electron energy. This influences wavelength and amplitude (thus penetration) of the X-rays produced. **Current** is usually 1 mA–1 A, and determines frequency of electron emission from the cathode. This affects the number of X-rays produced per second, which influences image resolution and total dose.

Geiger Muller tubes or scintillation counters can detect radiation. **Geiger Muller tubes** comprise an inert gas-filled chamber containing helium, neon or argon, with a high voltage applied across it. Radiation ionises the gas, allowing current to flow in proportion to the amount of radiation present. **Scintillation counters** use sodium or caesium iodide. Radiation striking these substances causes emission of electromagnetic waves of light (**scintillation**), which is detected by a photomultiplier tube and converted into a reading corresponding to the amount of radiation.

Applied science

What clinical uses are there for radiation?

Radiation can be used for diagnostic (imaging), therapeutic and sterilisation purposes.

Plain film radiographs and **computerised tomography** scans use X-rays, which penetrate body tissues and are sensed by a detector. Photographic paper consisting of silver halide was traditionally used, but newer digital X-rays use modified scintillation detectors or semi-conductors made of silicon or germanium. Semi-conductors are able to conduct a current when they absorb energy from X-rays; this enables a radiographic image to be produced when processed.

The gamma camera is a modified scintillation detector used to detect radioisotopes injected, inhaled or ingested. Thallium-201 has a half-life of 72 hours and emits 70–80 kV gamma rays. It is used for myocardial perfusion scanning, or detection of parathyroid tumours. Technetium-99 has a half-life of 6 hours and emits 140 kV gamma rays. It is used for functional imaging of organs including the brain, heart and kidneys. Its short half-life limits the radiation dose.

Positron emission tomography (PET) uses fluorine-18, which emits positrons. Fluorodeoxyglucose containing fluorine-18 is taken up by rapidly metabolising tissues. Positrons combine with nearby electrons, releasing two identifiable gamma rays in opposite directions. Detection of these rays can locate metabolically active cancer cells.

Rapidly dividing cancers are sensitive to ionising radiation. **Teletherapy** (external radiation) targets gamma rays released from cobalt-60 or X-rays at internal cancers. **Gamma knife** surgery is a sophisticated form of teletherapy where deep brain tumours can be specifically targeted by a halo array of emitters. This reduces the radiation dose to surrounding tissues. **Brachytherapy** (short-range radiation) uses implants of iridium-192 to target breast and prostate cancers. **Iodine-131** is administered intravenously and taken up by thyroid tumours, which it eliminates.

Sterilisation of medical equipment uses gamma and X-rays to destroy bacteria and viruses.

How can radiation exposure be minimised?

The Ionising Radiation (Medical Exposure) Regulations 2000 (IRMER) guidelines published by the Department of Health provide a framework to protect patients and staff. Recommendations include the following.

Time: minimising the time an ionising radiation source is used.

Distance: intensity of radiation is inversely proportional to the square root of the distance from the source.

Shielding: lead shields of 7 cm thickness will absorb 90% of gamma rays and X-rays. Staff members should wear aprons and thyroid shields, and susceptible parts of patients should be covered.

Storage: when not in use, radioisotopes must be stored in a locked safe environment, ideally lead- or concrete-lined. Warning symbols should be present on devices that contain radiation sources.

Employee dose limits: photographic film badges worn by employees working with radioactive sources detect the amount of radiation they are exposed to. This should not exceed 20 mSv per year.

Administration: **a radiation safety officer** must be employed by the trust. Their role is to:

- authorise the purchase of and inspect new equipment,
- oversee maintenance of equipment and storage of isotopes,
- produce guidelines for unintentional spillage of radioactive materials,
- train staff in appropriate conduct with radioactive materials,
- have emergency measures for misplaced radioactive sources.

Questions

Define atomic number and atomic mass.
What types of radiation are there?
What is the SI unit for X-ray dose received?
How are X-rays produced?

Index

Note: Locators in **bold type** refer to figures and tables.

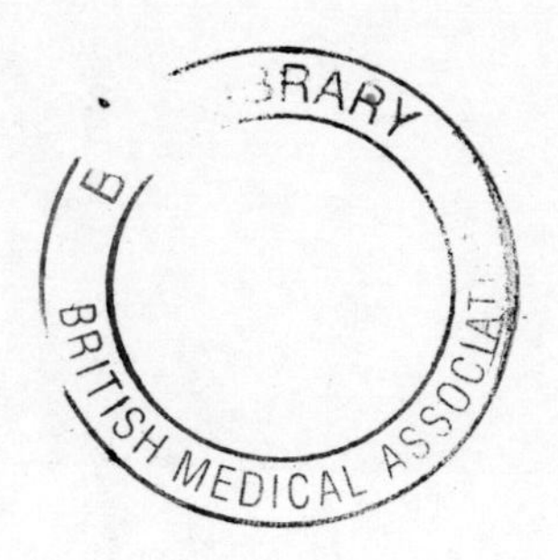